Green Energy and Technology

Climate change, environmental impact and the limited natural resources urge scientific research and novel technical solutions. The monograph series Green Energy and Technology serves as a publishing platform for scientific and technological approaches to "green"—i.e. environmentally friendly and sustainable—technologies. While a focus lies on energy and power supply, it also covers "green" solutions in industrial engineering and engineering design. Green Energy and Technology addresses researchers, advanced students, technical consultants as well as decision makers in industries and politics. Hence, the level of presentation spans from instructional to highly technical.

Indexed in Scopus.

More information about this series at http://www.springer.com/series/8059

A. P. Schaffarczyk

Introduction to Wind Turbine Aerodynamics

Second Edition

 Springer

A. P. Schaffarczyk
Mechanical Engineering Department
University of Applied Sciences
Kiel, Germany

ISSN 1865-3529   ISSN 1865-3537   (electronic)
Green Energy and Technology
ISBN 978-3-030-41030-8   ISBN 978-3-030-41028-5   (eBook)
https://doi.org/10.1007/978-3-030-41028-5

This Springer imprint is published by the registered company Springer Nature Switzerland AG
The registered company address is: Gewerbestrasse 11, 6330 Cham, Switzerland

*This book is dedicated to my wife Ulrike and
our children Claudia, Carola and Kai.*

Foreword

The modern development of wind power is a remarkable story of the combined effort of enthusiastic entrepreneurs and skilled engineers and scientists. Today, wind power forms the most rapidly advancing renewable energy resource with an annual growth rate of about 30%. Within the last 20 years the size of wind turbines have increased from a rotor diameter of about 30 m to 150 m, corresponding to an increase in power by a factor of more than 25. In the same period of time, the knowledge and scientific level of the aerodynamic research tools to develop optimally loaded rotor blades have increased dramatically. Today, wind turbine aerodynamics forms one of the research frontiers in modern aerodynamics. The aerodynamics of wind turbines concerns modeling and prediction of the aerodynamic forces on the solid structures of a wind turbine, and in particular on the rotor blades of the turbine. Aerodynamics is the most central discipline for predicting performance and loadings on wind turbines, and it is a prerequisite for the design, development, and optimization of wind turbines. From an outsider's point of view, aerodynamics of wind turbines may seem simple as compared to aerodynamics of, e.g., fixed-wing aircraft or helicopters. However, there are several added complexities. Most prominently, aerodynamic stall is always avoided for aircraft, whereas it is an intrinsic part of the wind turbines' operational envelope. Furthermore, wind turbines are subjected to atmospheric turbulence, wind shear from the atmospheric boundary layer, wind directions that change both in time and in space, and effects from the wake of neighboring wind turbines. *Introduction to Wind Turbine Aerodynamics*, written by an experienced teacher and researcher in the field, provides a comprehensive introduction to the subject of wind turbine aerodynamics. It is divided into ten chapters, each giving a self-contained introduction to specific subjects within the field of wind turbine aerodynamics. The book begins with an overview of different types of wind turbines, followed by a comprehensive introduction to fluid mechanics, including different forms of the basic fluid mechanics equations and topics such as potential flow theory, boundary layer theory, and turbulence. The momentum theory, which still forms the backbone in the rotor design of wind turbines, is presented in a separate chapter, which, in a compressed form, describes the basic momentum theory as well as the important blade-element momentum theory. State-of-the-art advanced aerodynamic models are also presented. This includes vortex models as well as computational fluid dynamics (CFD) techniques and an introduction to the most common turbulence models. The last part of the book gives an overview of experiments in wind turbine aerodynamics and the impact of aerodynamics on blade design. Each chapter contains some homework problems and concludes with a bibliography. *Introduction to Wind Turbine Aerodynamics* is a highly topical and timely textbook of great value for undergraduate students as well as for trained engineers who wish to get a fast introduction into the many aspects of the amazing world of wind turbine aerodynamics.

Jens Nørkær Sørensen, Technical University of Denmark, Kongens Lyngby, Denmark.

Copenhagen, Denmark   J. N. Sørensen
April 2014

Preface to Second Edition

About 5 years have passed since the first edition appeared. Thanks to Springer (esp. Dr. Baumann) a second edition was prepared reflecting the progress made during this time. As predicted by Ben Backwell [1] from Chap. 1 the market sees some manufacturers to disappear and a lot of merging. Nevertheless Renewables will account for about 200 GW **new** installed capacity in 2019 with 100 GW from Photovoltaics, 55 GW from onshore and about 5 GW from offshore wind energy.

To my opinion progress in wind turbine aerodynamics was most notable for

- a 12 MW machine with 220-m rotor diameter is now in operation,
- on-shore machines of 3 MW are now state of the art,
- much more investigation on transition on wind turbine blades,
- IEAwind Annex 29 was extended to *Aerodynamics or Analysis of Aerodynamic Measurements*.

Statistics were updated whenever reasonable data was available. Values in Table 5.1 were corrected (My thanks go to Emmanuel Brandlard, NREL). A new Sect. 4.6 about artificial turbulence especially for Large Eddy Simulation was prepared by B. Lobo.

Kiel, Germany
December 2019

A. P. Schaffarczyk

Reference

1. Backwell B (2018) Wind power—The struggle for control of a new global industry, 2nd edn. Earthscan, Oxon

Preface to First Edition

Wind energy has flourished over the last 20 years. At the time of this writing, approximately 40 GW of wind power production capacity, equivalent to 80 Billion US $ in investment, is added to the worldwide energy portfolio annually.

About 2,40,000 (IEA Wind 2012 annual report) engineers are working in this field, and many of them are confronted with challenges related to aerodynamics. Even special academic programs (one of them being the M.Sc. Program in Wind Engineering at the consortium of universities in Schleswig-Holstein, the north-ernmost province of Germany) now are educating young people in this field. From personal experience, the author knows how difficult it is to navigate through the jungle of scientific literature without being injured before having learned all the basics. This textbook intends to help young and old interested people to gain easier access to the basics of the fluid mechanics of wind turbines.

The scope of this book is to take the reader on an interesting journey from the basics of fluid mechanics to the design of wind turbine blades. The reader's basic knowledge of mathematics (Vector calculus) seems to be necessary for under-standing, whereas fluid mechanics will be explained in what amounts to a crash course. But there should be no mistake: fluid mechanics being *non-linear*, even as an applied science, is a complicated affair and can only be learned and understood by one's *own* problem solving. This is especially true when investigating *turbulent* flow which is of utmost importance when investigating mechanical loads on wind turbine parts. Therefore, the estimated amount of effort necessary for solving our selected problems at the end of each chapter ranges from 10 min to 10 months.

The author will try to give the beginning reader a glimpse of established knowledge about the aerodynamics of wind turbines. For the more familiar reader, this book will facilitate the discovery of (and hopefully understanding of) relevant advanced papers. In any event, all readers will glean a deeper understanding which will allow them to perform analyses either by hand or by using computer codes.

The structure is as follows: an introduction is given to inform the reader about the specific relevance of wind turbine aerodynamics, while touching on some interesting historical examples. Then a chapter on installed windmills and wind turbines is added to describe the basics of mainstream and unconventional technology. After that we are ready to tackle our general framework: the laws of fluid mechanics. I will try to limit the contents to those topics that are closely related to wind, understood as turbulent flow in the lower (boundary layer) part of the atmosphere. Because the rotor swept area now easily reaches several *hectares* ($= 10^4$ m^2) something must be said about real-world inflow conditions, which are not constant from a spatial or temporal point of view. Chapter 5 is the core of the book and explains the various versions of *Momentum Theory* as well as its limitations, followed by a description of applications of its counterpart: *Vortex Theory*. After exhausting these classical methods, we then have to present the modern approach of solving the differential equations called *Computational Fluid Dynamics* in some detail. Many efforts have been expended over the last 15 years to bridge the gap from older, sometimes called simpler, theories to actual measurements using these computational methods. This leads directly to a transition in the next chapter to a discussion of free-field and wind tunnel measurements. After that we will try to get even closer to practical, real things. This means that we try to give examples of actual blade shapes from industry. We conclude with summarizing remarks and an overview of possible future developments.

As this text grew out of my lectures on wind turbine aerodynamics in the first year of our international Master of Science Program, it is targeted toward this student audience. Nevertheless, I have tried as much as possible to make it valuable and informative for a much broader group of readers.

Kiel, Germany   A. P. Schaffarczyk
April 2014

Acknowledgements

I want to thank several people for introducing me to wind turbine aerodynamics: First of all Sönke Siegfriedsen from aerodyn Energiesysteme GmbH, Rendsburg and Gerard Schepers, ECN, the energy research center of the Netherlands, who gave me a first chance in 1997. My very first contact with wind energy was actually somewhat earlier (1993) during supervising a Diploma-Thesis which is now summarized as the *strange farmer* problem (see Problem 5.1). Participation in several IEAwind tasks (11, 20 and 29) gave me the opportunity to interact with leading international experts. State of Schleswig-Holstein, Federal German and European funding helped contribute to important projects and finally an International M.Sc. program in *Wind Engineering* was established in 2008 in which now more than 160 young people from all around the world (hopefully) learned about wind energy technology.

Mr. Lippert prepared most of the graphs and Mr. John Thayer brought the text into more correct English. Any errors, however are in the responsibility of the author only.

Contents

Nomenclature

$< \bullet >$	Averaging
u_τ	Friction velocity
α	Angle-of-attack
α, β	Component counting tensor indices
β	Camber of Joukovski airfoil
$\bullet^+$	Dimensionless turbulent quantities
$\delta(\mathbf{u}r)$	Structure function
$\delta^\star$	Displacement thickness of boundary layer
$\dot{m}$	Mass flow, unit kg/sec
ℓ	Arbitrary extension of profile
ℓmix	Prandtl's mixing length
ε	Dissipation
η	Dynamic viscosity
η	Kolmogorov's length scale
η	Similarity parameter: Normalized distance from wall
Γ	Circulation
$\Gamma(x)$	Gamma function
$\gamma(x)$	Linear distributed vorticity
κ	Adiabatic exponent
κ	Vortex strength of a vortex sheet
Λ	Pressure parameter in wedge-flow
λ	Parameter of Joukovski's transformation
λ	Tip-speed-ratio
λ_T^2	Taylor's microscale
$\mathscr{F}$	Complex (2D) force
μ	Aspect ratio
∇	3D vector of Cartesian partial derivatives
v	Kinematic viscosity
v_t	Turbulent viscosity
Ω	Rotational velocity of turbine (rad/sec)

ω, ω_1	Angular wake velocity at disk and far downstream
Φ_{ij}	Spectral tensor
ψ	Stream function
$\psi_m(z/L)$	Function specifying the state of the ABL
ρ	Density of air, in most cases about 1.2 kg/m^3
$\rho(s)$	Autocorrelation
σ	Solidity of blade
σ_v	Variance of wind-speed measurement
τ	Shear stress
τ_w	Wall shear stress
$\tau_{\alpha\beta}$	Reynold's stress tensor
θ	Momentum thickness of boundary layer
Φ	Velocity potential
φ	Flow angle
ζ	Complex quantity
A	Constant in Hill's spherical vortex
a	Axial induction factor
a'	Tangential induction factor
A, k	Scaling factors in Weibull distribution
A_n	Fourier coefficients in thin airfoil theory
A_r	The reference area for capturing wind
c	Chord of Joukovski airfoil
c	Velocity of sound
C^∞	Hilbert space of infinitely differentiable functions
c_D	Drag coefficient
C_f	Friction coefficient
C_K	Kolmogorov's constant
c_L	Lift coefficient
c_N	Normal force coefficient
c_P	Fraction of the power of the wind converted to useful power
c_p	Specific heat at constant pressure
c_{D+}	Drag coefficient in wind direction
c_{D-}	Drag coefficient opposite to wind direction
c_{tan}	Tangential force coefficient
$D = 2R_{tip}$	Rotor diameter
D	Drag force in wind direction
d	Spacing of surrogate model, see Eq. (A.66)
$E(k)$	Elliptic integral
F	Prandtl's tip (loss) factor
$f(r)$	Function in Eq. (3.107)
$F(z)$	Complex 2D velocity potential
G	Goldstein function
g	Earth's acceleration
$g(r)$	Function in Eq. (3.107)

H	Shape parameter
I	Turbulence intensity
J	Advance ratio
K	Drag coefficient for wind vehicle
k	Kinetic turbulent energy
k	Von Karman's constant
$K(k)$	Elliptic integral
L	Integral correlation length
L	Lift force perpendicular to inflow velocity
M	Torque of rotor
N	Amplification exponent
P	Probability
p_0	Ambient pressure
P_{Wind}	The power of wind
Q	Torque on low-speed shaft
$Q_{\alpha\beta}(\mathbf{r})$	Correlation tensor
r_2, r_3	Distance of regarded streamline form center at disk and far downstream
R_i	Ideal gas constant
Re	Reynolds number: Local ratio between viscous and inertia forces
Re_x	Reynolds number with respect to a running length
$S(\mathbf{r})$	Source function for pressure from Poisson's equation, Eq. (3.60)
S_n	Structure functions
T	Absolute temperature in K(elvin)
t	Thickness of Joukovski airfoil
T_R	Return time of maximum wind speed
Ti	Turbulence intensity
u	Speed of moving vehicle
u'	Fluctuating part of velocity
$U = <u(t)>$	Averaged velocity
u_1	Axial velocity far up-stream
u_2	Axial velocity at disk
u_3	Axial velocity far down-stream
u_e	Velocity at the edge of the boundary layer
U_S	Normalized gust wind-speed
v	Wind velocity
z	Vertical coordinate, positive upwards
z_0	Roughness height, location where wind is zero
κ	Wave vector
ω	Local vorticity
σ	Cauchy stress tensor
$\mathbf{A}$	Vector potential
$\mathbf{a}$	Acceleration in an Eulerian (Earth fixed) frame of reference
$\mathbf{b}$	Unit bi-normal vector of a space curve

D	Deviation (deformation) tensor
F	Force in Newton's second law
H	Bernoulli's constant or specific enthalpy
n	Unit normal vector of a space curve
T	Thrust—force in wind direction
t	Unit tangent vector of a space curve

List of Figures

Chapter 1
Introduction

1.1 The Meaning of Wind Turbine Aerodynamics

Wind energy has been used for a long time. Chap. 1 of [2] gives a short account of this history and describes the first scientific approaches. It is immediately seen that even this applied branch of fluid mechanics suffers from a general malady: Prior to the twentieth century, mathematical descriptions and practical needs were far removed from one another. Impressive reports about the history of fluid mechanics can be found in [7, 13, 18], in Chap. 3. There it is shown how this dilemma was resolved at least partly. We will follow this approach and will try to combine applied theory with practical needs as far as possible.

How much power is in the wind? Imagine an area—not necessarily of circular shape—of A_r. The temporally and spatially constant wind may have velocity v perpendicular to this area, see Fig. 1.1. Within a short time dt this area moves by $ds = v \cdot dt$. This then gives a volume $dV = A_r \cdot ds$. Assuming constant density $\rho = dm/dV$ we have a mass of $dm = \rho \cdot dV = \rho A_r v \cdot dt$ in this volume. All velocities are equal to v, so the kinetic energy content $dE_{kin} = \frac{1}{2}dm \cdot v^2$ increases as the volume linearly with time. Defining power as usual by $P := dE/dt$ we summarize

$$P_{wind} = \frac{dE}{dt} = \frac{1}{2} \cdot \rho \cdot A_r \cdot v^3 . \tag{1.1}$$

We may transfer this to an energy density (W/m^2) by dividing by the reference area A_r. Using an estimated air-density of 1.2 kg/m^3 we see that for $v = 12.5$ m/s this power-density matches the so-called solar constant of about 1.36 kW/m^2. This energy-flux (J/m$^2 \cdot$ s) is supplied by the sun at a distance of one astronomical unit (the distance from the sun to the earth) and may be regarded as the natural reference unit of all renewable power resources.

As $p \cdot \dot{V}$ (static pressure times volume flow) also gives power, it is sometimes argued (in Sect. 5.2) that by decreasing ambient pressure (around 1013 hPa) up to far downstream even more power my be extracted. We will come back to this idea later.

© Springer Nature Switzerland AG 2020
A. P. Schaffarczyk, *Introduction to Wind Turbine Aerodynamics*,
Green Energy and Technology, https://doi.org/10.1007/978-3-030-41028-5_1

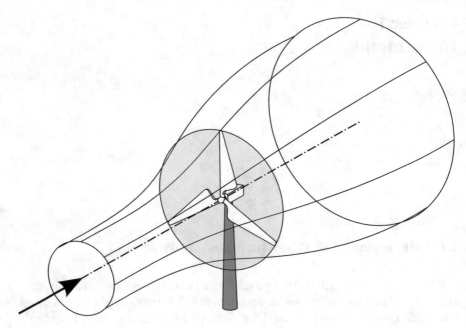

Fig. 1.1 Wind turbine with slip stream

Everybody knows that wind is NOT constant in time, and therefore statistics must be involved. In the simplest form, a family of two-parameter functions is used:

$$P(v; A, k) := \frac{k}{A} \left(\frac{v}{A}\right)^{k-1} \exp\left(-\frac{v}{A}\right)^{k}. \tag{1.2}$$

It is a commonly used probability distribution attributed to Weibull.[1] A is connected with the annual-averaged mean velocity $\bar{v}$ and k is the shape factor with values between 1 and 4. For most practical applications:

$$\bar{v} := \int_0^\infty P(v; A, k) v \cdot dv \approx A \left(0.568 + \frac{0.434}{k}\right)^{\frac{1}{k}} \tag{1.3}$$

Figure 1.2 shows some examples of Weibull distributions with shape parameters varying from 1.25 to 3.00 In Figs. 1.3 and 1.4, we compare fitted (by the least square algorithm) distributions with a measured distribution. We see that the 2008 data fits rather well whereas (note the logarithmic y-scale!) 2011 is much worse. Thus, unfortunately we must conclude that the distributions derived from annual data are not constant in time. This has long been known in wind energy as the *Wind Index Problem* . It means that a long term (e.g., 50 year) average may not be constant over time. Figure 1.5 shows that even over a period of about 10 years, large fluctuations

---

[1]A Weibull distribution with k = 2 is called a *Rayleigh distribution*.

Fig. 1.2 Weibull
distributions

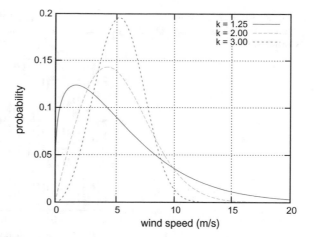

Fig. 1.3 Comparison with
measured data from Kiel,
Germany, 2008

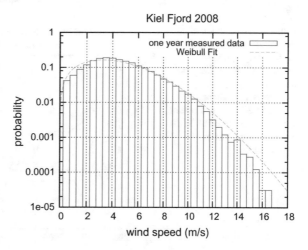

of about 20% are possible. With Eq. (1.2) some useful limits of annual yield ($=$
useful mechanical work $=$ power $\times$ time $=$ P $\times$ 24 $\times$ 365.25) can be given, under
the assumption that only a fraction $0 \leq c_P \leq 1^2$ of Eq. (1.1) can be given. Using

$$m_n := \int_0^{\infty} P(v; A, k)v^n \cdot dv = A^n \cdot \Gamma \left(1 + \frac{n}{k}\right) \tag{1.4}$$

and

$$P := c_P \cdot \frac{\rho}{2} A_r \cdot v^3, \tag{1.5}$$

we have for an annual averaged power in the case of k $=$ 2 (Rayleigh distribution):

$^2 c_P$ is the efficiency of a wind turbine.

Fig. 1.4 Comparison with measured data from Kiel, Germany, 2011

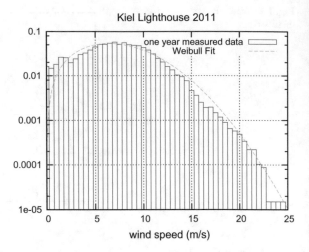

Fig. 1.5 Wind Index of Schleswig-Holstein applied to the energy yield of a research wind turbine. 1 = nominal output, Note that there was an exchange of blades in 2011. Rated power was partly throttled in 2018 because of extension of operating permit

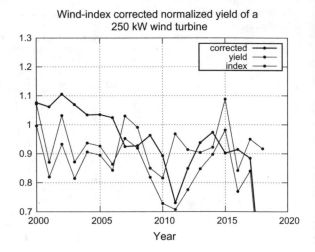

$$\bar{P} := c_P \cdot \frac{\rho}{2} A_r \cdot R^3 \cdot \Gamma(2.5) = 0.92 \cdot c_P \cdot P \,. \tag{1.6}$$

Sometimes this efficiency number is translated into *full-load-hours* which must be less than 8760 h per year, or into capacity factors. In a later chapter, we will come back and see how state-of-the-art wind turbines fit into this picture.

Of course the question arises how we may justify or even derive special types of wind occurrence probability distributions. A very simple rationale starts from the *central limit theorem* of probability theory together with the basic rules for transformation of stochastic variables. van Kampen [5] describes it in a somewhat exaggerated manner:

The entire theory of probability is nothing but transforming variables.

Fig. 1.6 Wind energy
market 2000–2018, partly
estimated, *Sources* BTM
Consult, Wind Power
Monthly and Global Wind
Energy Council

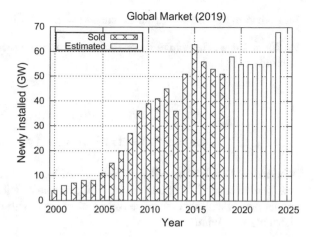

We will come back to this interesting question in the context of the statistical theory
of turbulence, Sect. 3.7.

There is one concluding remark about comparing the economical impact of wind-
turbine blade aerodynamics to helicopters [6] and ship propellers [3]. Assuming a
worldwide annual addition of 50 GW of installed wind power, the resulting revenue
for just the blades is about 12.5 Billion Euros. About one thousand helicopters (80%
of them for military purposes) [6] may produce approximately 50 Billion Euros with
an unknown smaller amount for just the rotor—estimated less than 10%. Assuming
also 1000 new ships are built every year, each of them having a propeller of about
100 (metrics) tons of mass, the resulting revenue is 1.5 Billion Euros.

All in all, we see that in less than 25 years, wind energy rotor aerodynamics has
exceeded the economic significance of other longer established industries involving
aerodynamics or hydrodynamics (Fig. 1.6).

1.2   Problems

Problem 1.1 Calculate the instantaneous power-density for v-wind = 6, 8, 10, and
12 m/s. Compare with the annual averaged values (Rayleigh distributed, k = 2, then
$A = \bar{v}$) of 4, 5, 6, and 7 m/s. Take density of air to $\rho = 1.26$ kg/m^3

Problem 1.2 Assume $c_P = 0.58$, $c_P = 0.5$, and $c_P = 0.4$. Calculate the corresponding annual averaged $\bar{c}_P$

Problem 1.3 Assume that a 2D (velocity) vector has a Gaussian 2-parameter distribution

$$G(x; m, \sigma) := \frac{1}{\sigma\sqrt{2\pi}} \exp{-\frac{1}{2}\left(\frac{x-m}{\sigma}\right)^2} \tag{1.7}$$

for both components $\mathbf{v} = (u,v)$. Calculate the distribution of length $y = \sqrt{u^2 + v^2}$.

Problem 1.4 A German reference site is defined as a location with vertical variation of wind speed by

$$u(z) = u_r \cdot \frac{log(z/z_0)}{log(z_r/z_0)}, \tag{1.8}$$

$$u_r = 5.5 \text{ m/s at } z_r = 30 \text{ m and } z_0 = 0, 1 \text{ m}. \tag{1.9}$$

A historical Dutch windmill with $D = 21$ m, $z_{hub} = 17.5$ m is quoted to have a (5-year's) reference yield of 248 MWh. Calculate an efficiency by relating the average annual power with that the same rotor would have with $c_P = 16/27$ and a Rayleigh distribution of wind.

References

1. Backwell B (2018) Wind Power - The struggle for control of a new global industry, 2nd edn. Earthscan, Oxon
2. Beurskens J (2014) History of wind energy, chapter 1 of: understanding wind power technology, Schaffarczyk AP (ed). Wiley Ltd, Chichester
3. Breslin JP, Andersen P (1996) Hydrodynamics of ship propellers. Cambridge University Press, Cambridge
4. Carlin PW (1996) Analytic expressions for maximum wind turbine average power in a Rayleigh wind regime, NREL/CP-440-21671, Golden, CO, USA
5. van Kampen NG (2007) Stochastic processes in physics and chemistry, 3rd edn. Elsevier, Amsterdam
6. Leishman JG (2006) Principles of helicopter aerodynamics, 2nd edn. Cambridge University Press, Cambridge
7. McLean D (2013) Understanding aerodynamics. Boeing, Wiley Ltd, Chichester
8. Schmitz S (2020) Aerodynamics of wind turbines. Wiley Ltd, Hoboken

Chapter 2
Types of Wind Turbines

*Ich halte dafür, daß das einzige Ziel der Wissenschaft darin
besteht, die Mühseligkeit der menschlichen Existenz zu
erleichtern (B. Brecht, Life of Galileo, 1941). [4] (Presumably
for the principle that science's sole aim must be to lighten the
burden of human existence.)*

Equation (1.5) from Chap. 1 may also be used to define an efficiency or *power coefficient* $0 \leq c_P \leq 1$:

$$c_P = \frac{P}{\frac{\rho}{2} A_r \cdot v^3} \, . \tag{2.1}$$

Wind turbine aerodynamic analysis frequently involves the derivation of useful equations and numbers for this quantity. Most (in fact: almost all) wind turbines use *rotors* which produce *torque or* moment of force

$$M = P/\omega \tag{2.2}$$

with $\omega = RPM \cdot \pi/30$ the angular velocity. Comparing tip speed $v_{Tip} = \omega \cdot R_{Tip}$ and wind speed we have

$$\lambda = v_{Tip}/v_{wind} = \frac{\omega \cdot R_{Tip}}{v_{wind}} \tag{2.3}$$

the (TSR). Figure 2.1, sometimes called the *map of wind turbines*, gives an overview of efficiencies as a function of nondimensional RPM. The numbers are estimated efficiencies only. A few remarks are germane to the discussion. From theory (Betz, Glauert) there are clear efficiency limits, but no theoretical maximum TSR. In

© Springer Nature Switzerland AG 2020

A. P. Schaffarczyk, *Introduction to Wind Turbine Aerodynamics*,
Green Energy and Technology, https://doi.org/10.1007/978-3-030-41028-5_2

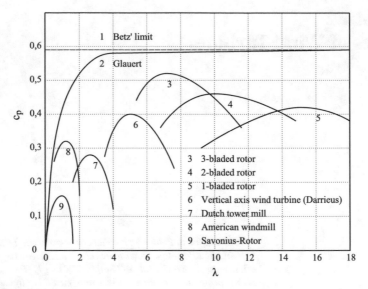

Fig. 2.1 Map of wind turbines

contrast, the semi-empirical curves for each type of wind turbine have a clearly defined maximum efficiency value.

2.1   Historical and State-of-the-Art Horizontal-Axis Wind Turbines (HAWT)

The wind energy community is very proud of its long history. Some aspects of this history are presented in [2, 23]. Apparently the oldest one [23]. It is the so-called *Persian* windmill (Fig. 2.2). It was first described around 900 AD and is viewed from our system of classification (see Sect. 2.6) as a drag-driven windmill with a vertical axis of rotation.

Somewhat later the *Dutch* windmill appeared as the famous *Windmill Psalter* of 1279 [23], see Fig. 2.3. This represented a milestone in technological development: the axis of rotation changed from vertical to horizontal. But also from the point of view of aerodynamics, the Dutch concept began the slow movement toward another technological development: lift replacing drag (Fig. 2.4). These two types of forces simply refer to forces perpendicular and in-line with the direction of flow. The reason this is not trivial stems from *D'Alembert's* Paradox:

Theorem 2.1 *There are no forces on a solid body in an ideal flow regime.*

We will continue this discussion further in Chap. 3. These very early concepts survived for quite a few centuries. Only with the emergence of electrical generators and

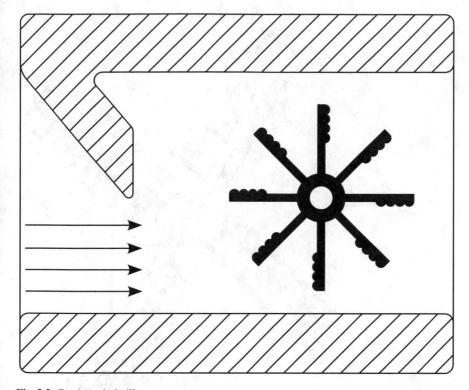

Fig. 2.2 Persian windmill

airplanes were these new technologies adapted, in the course of a few decades, to what is now called the standard horizontal-axis wind turbine:

- horizontal axis of rotation,
- three bladed,
- driving forces mainly from lift,
- upwind arrangement of rotor; tower downwind,
- variable speed/constant TSR operation
- pitch control after rated power is reached

2.2   Nonstandard HAWTs

With this glimpse of what a *standard* wind turbine should be, everything else is *nonstandard*:

- **no** horizontal axis of rotation,
- number of blades other than three (one, two, or more than three),

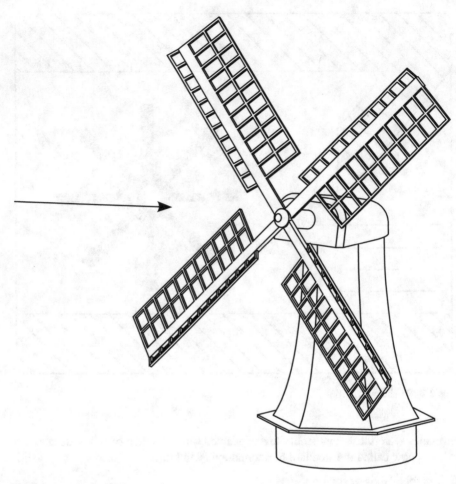

Fig. 2.3 Dutch windmill

Fig. 2.4 Lift and drag forces
on an aerodynamic section

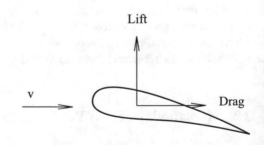

- drag forces play an important role,
- downwind arrangement of rotor; tower upwind,
- constant speed operation,
- so-called **stall** control after rated power is reached.

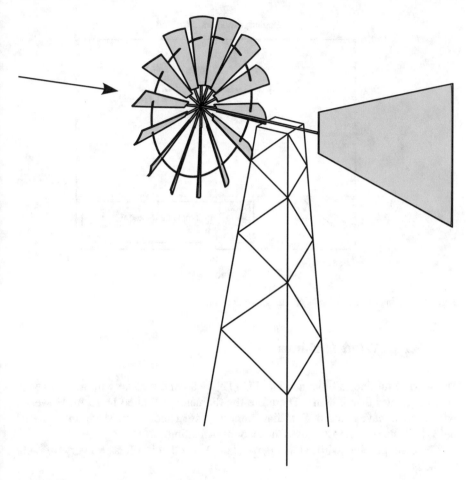

Fig. 2.5 American or Western windmill

From these characteristics, we may derive a large number of different designs. Only a few of them became popular enough to acquire their own names:

The *American* or *Western-type* turbine [23] Chap. 1 uses a very high (10–50) number of blades which in most cases are flat plates with a small angle between plane of rotation and chord. These turbines were used mainly in the second half of the nineteenth century, see Fig. 2.5

The *Danish Way* of extracting wind energy [18] used most of the now classical properties with a fixed-pitch blade arrangement and a constant-RPM operation mode. The development of this design philosophy started in the 1940s and died off slowly in the 1990s.

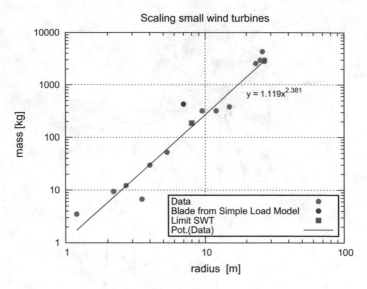

Fig. 2.6 Scaling of blade masses of smaller wind turbines

2.3   Small Wind Turbines

Small wind turbines are defined by IEC [12] as a wind turbine with a rotor swept area no greater than $200\,\mathrm{m^2}$. Therefore the diameter is limited to 16 m. However, most of them have much smaller diameters starting at about 1 m. More can be found in [37], Chap. 9. Figure 2.6 gives an account of scaling.

The main problem with safety approval of [12] is that it offers two very different methods:

- the usual aeroelastic simulation modeling,
- a simplified load model.

The first one implies the same amount of work as for a state-of-the-art turbine and is not economical in most cases. The second procedure is much easier (see [37] Chap. 9) but at the expense of exceedingly high safety factors. As an example the required blade-mass for the simple load model is shown in Fig. 2.6. The mass required by the simple load model has to be more than 300 kg, compared to only 120 kg for a blade designed without these high safety factors.

2.4   Vertical Axis Wind Turbines

As was explained earlier, vertical axis windmills and the subsequent vertical axis wind turbines seem to be older than those with a horizontal axis of rotation. Mainly

Savonius-Rotor · · · · · · · · · · Darrieus-Rotor · · · · · · · · · · H-Darrieus-Rotor

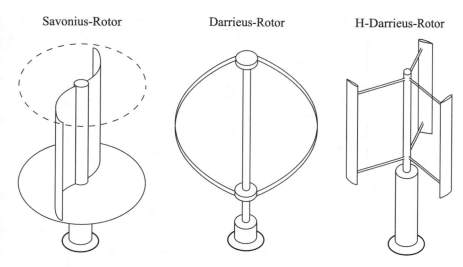

Fig. 2.7 Types of vertical axis wind turbines

due to the inventions of Darrieus [7] and Savonius, [40][1] interest in these vertical turbines was renewed in the early twentieth century. Then, after the first so-called oil crisis in the beginning of the 1970s many US [2, 6, 11, 19] and German [29–32, 34, 35, 37] vertical turbine development efforts were undertaken which lasted until the 1990s, while HAWTs also progressed. A summary may be found in [20]. Now, after some 20 years of dormancy, interest in VAWTs seems to be returning slowly, see for example [8] (Fig. 2.7).

One of the big advantages is the independence of directional change in wind. See Fig. 2.8 for a typical distribution of wind direction at a typical site. Also heavy components may be installed close to the ground, as as illustrated in Fig. 2.9. The largest VAWT manufactured so far was the so-called Éole-C made in Canada. Its height was about 100 m, the rotating mass was 880 metric tons, and the rated power was supposed to be 4 MW. Unfortunately due to severe vibration problems, the rotational speed was limited to such low values that only 2 MW was reached [20].

It has be to noted that the flow mechanics behind these designs (see Problems 6.1 and 6.2 in Chap. 6) are much more interesting—but also more complicated—due to fact that the sphere of the influence of the rotor is modeled as a real volume and not a 2D disk as is assumed for the HAWT actuator disk model. Therefore at least for each half-revolution, the blades are operating in a wake, meaning that load fluctuations have a much greater influence on the blades. The resulting so-called aerodynamic fatigue loads for a VAWT rotor blade are much higher and are one of the reasons that VAWTs are much more prone to earlier failure of components—mostly at joints— than are HAWTs.

---

[1] A more detailed discussion can be found in Sect. 2.6.

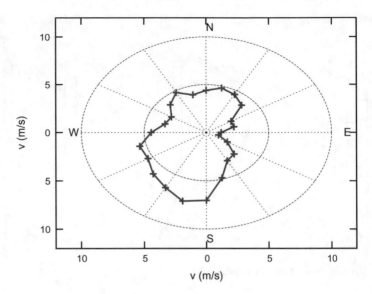

Fig. 2.8 Typical wind direction distribution near Hamburg at 10 m height

One way out of this difficulty is the so-called *Gyro-Mill* or articulated VAWT [42], where the pitch angle (the angle of the chord line in relation to the circumference) is changed periodically so that—in an ideal case—the driving force remains almost constant (Fig. 2.10).

2.5   Diffuser Augmented Wind Turbines

As we have seen, the power contained in the wind is proportional to the swept area. An obvious extension of this concept is to look for *wind-concentrating* devices resembling a cone or funnel, (see Fig. 2.11). Such devices are very common in wind turbines and are called draft tubes or suction tubes. From first principles of fluid mechanics, the **exit** area of such a device has to be larger than the inflow area. At a glance this is clearly counterintuitive. Then, by closer inspection of the basic laws of conservation of mass and energy—called *Bernoulli's* law—it follows that an increase of mass-flow proportional to the area ratio A_{exit}/A_{inflow} is possible if the flow follows the contour of the cone (Fig. 2.12).

Unfortunately nature is not that generous. The increase in diameter has to be very moderate. To be more precise: opening-angles less than 10° have to be used to avoid what is called flow separation. It immediately follows that to have reasonable area-ratios, we have to use very long diffusers with very large weights. The engineering task then is to find a reasonable compromise—if possible at all. Serious work started

Fig. 2.9 Dornier's Eole-D, height $= 20$ m, $P_{rated} = 50$ kW, Reproduced by permission of EnBW Windkraftprojekte GmbH, Stuttgart, Germany

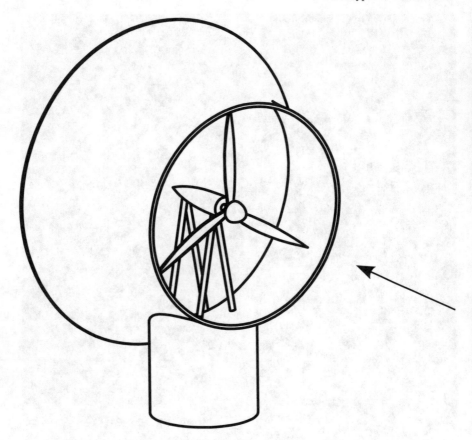

Fig. 2.10 Example of a DAWT, wind from right

in 1956 by Liley and Rainbird [14] and efforts up to 2007 are summarized by van Bussel [5]. Some applications to small wind turbines may be found at [39].

2.6 Drag-Driven Turbines

Drag was defined earlier as a force on a structure subjected to a stream of air **in line** with the flow, see Fig. 2.4. We, therefore, define a simple number, the *drag coefficient*:

$$c_D := \frac{D}{\frac{\rho}{2} \cdot A_r \cdot v^2} \, . \tag{2.4}$$

To imagine this, we consider at first a simply translating sail of velocity u and Area $A = c \cdot \ell$ and v the wind velocity as usual. The power $P = D \cdot u$, is using the

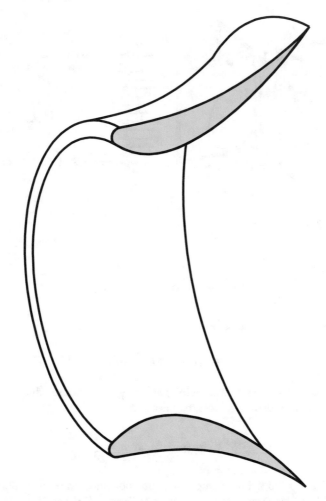

Fig. 2.11 Cross section of a diffuser or ring-wing

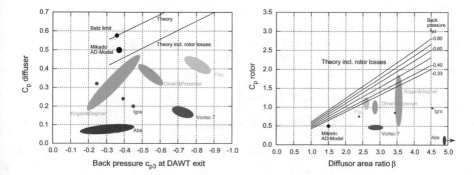

Fig. 2.12 Collected performance data for diffusers, from van Bussel (2007)

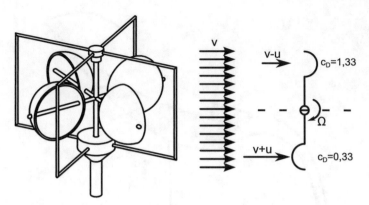

Fig. 2.13 Drag-driven turbines

expressions above:

$$P = \frac{\rho}{2}(v-u)^2 \cdot c_D \cdot c \cdot \ell \cdot u \ . \tag{2.5}$$

Now using $P_{wind} = \frac{\rho}{2}v^3$) we have

$$c_P = c_D \left(1 - u/v\right)^2 \left(u/v\right) . \tag{2.6}$$

Setting $a := u/v$ and solving for a by making $\frac{dc_P}{da} = 0$, we see that the maximum power of such a *drag-driven vehicle* may not exceed

$$c_P^{max,D} = \frac{4}{27} \cdot c_D \ \text{at} \ a = \frac{1}{3} \tag{2.7}$$

$c_P^{max,D} \approx 0.3$ if **u = 1/3 v**. For example, a sailing boat at wind force 7 (Beaufort $\approx$30 knots $\approx$16.2 m/s) may not travel faster than 10 Knots. If it has 30 m^2 sail area the maximum power will be P = 24 kW.

The next step is to discuss the *Persian windmill* or the closely related anemometer, (see Fig. 2.13). The idea is to use a specially shaped body (semi-sphere) which has different c_D when blown from one side or the other. A common pair of values for c_D is $c_{D+} = 1.33$ and $c_{D-} = 0.33$. We then arrive at

$$F_+ = c_{D+} \cdot \frac{\rho}{2} A_r \left(u - \Omega r\right), \tag{2.8}$$

$$F_- = c_{D-} \cdot \frac{\rho}{2} A_r \left(u + \Omega r\right), \ \text{and finally} \tag{2.9}$$

$$c_P = (F_+ - F_-) \cdot v/A_r = \lambda \left(c_1 - c_2 \cdot \lambda + c_1 \cdot \lambda^2\right) \tag{2.10}$$

with $c_1 = c_{D+} - c_{D-}$ and $c_2 = 2 \cdot (c_{D+} + c_{D-})$. Figure 2.14 shows then a very small efficiency at very low tip speed ratios ($c_{D+} = 1.33$ and $c_{D-} = 0.33$).

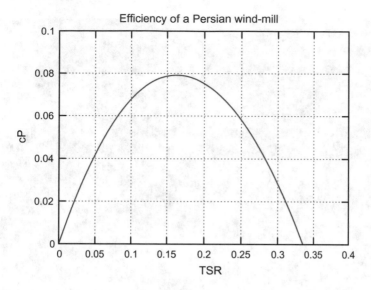

Fig. 2.14 Performance curve of an anemometer

2.7   Counter-Rotating Wind Turbines

F some ship propellers and helicopters, there have been efforts to use to rotors rotating in opposite direction to improve the efficiency of the whole system. Unlike propellers and helicopters, swirl losses (see Sect. 5.2) appear to be small in wind turbines. Nevertheless many investigations of swirl losses [10, 22] have been undertaken. Figure 2.15 shows a model wind turbine which was evaluated during a wind tunnel experiment in 2009 in Geneva. Clearly an improvement of about 10% was seen in performance but somewhat obscured by a rather large and uncertain blockage correction, which is of utmost importance to consider when comparing wind tunnel experiments with a freely expanding wake (Fig. 2.16).

From a theoretical point of view this type of turbine is very interesting, because a swirl component does not have to be included. We will come back to this point in Chap. 6.

2.8   Airborne Wind Turbines

The [1, 15] mechanism for driving airborne devices is somewhat similar to a simple translating device driven by an lifting airfoil (compare to Fig. 2.17).

Geometry is 2D and let's assume that wind and driving direction are perpendicular.

Power is $P = N \cdot v$ with N normal force alos perpendicular to wind and parallel to v. U and v may be completed for an triangle, N consist of

Fig. 2.15 Model of a counter-rotating wind turbine in a wind tunnel, diameter = 800 mm. Reproduced with permission of HEIG-VD, Iverdon-les-Bains, Switzerland

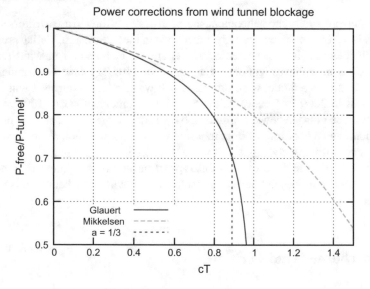

Fig. 2.16 Blockage corrections from Glauert [25] in Chap. 3 and Mikkelsen [16]

Fig. 2.17 Forces and
velocities for purely
translating drag device (top)
and a translating device
using lift (bottom)

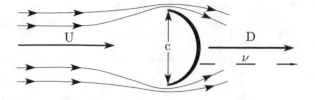

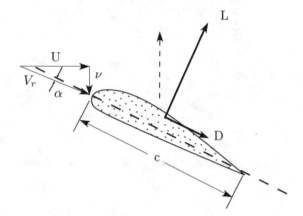

$$N = N_L + N_D = L \cdot cos(\varphi) - D \cdot sin(\varphi). \tag{2.11}$$

Some simple algebra gives

$$c_P = a \left(\sqrt{1+a^2}\right) \cdot (C_L - C_D \cdot a) \tag{2.12}$$

Maximum power coefficient (see Fig. 2.18) now is

$$c_P^{max,L} = \frac{2}{9}C_L \left(\frac{C_L}{C_D}\right) \sqrt{1 + \frac{4}{9}\left(\frac{C_L}{C_D}\right)^2} .. \tag{2.13}$$

Note:

- maximum velocity may be **larger as wind speed**—reached at

$$a = v/U = \frac{2}{3}\frac{C_L}{C_D} . \tag{2.14}$$

- There is a (not negligible) force T parallel to wind which must be compensated.

Figure 2.18 gives an example for $C_D = 2$, $C_L = 1$, and $C_L/C_D = 10$.

It has to be noted that [1] condens these and the investigation of [15] to

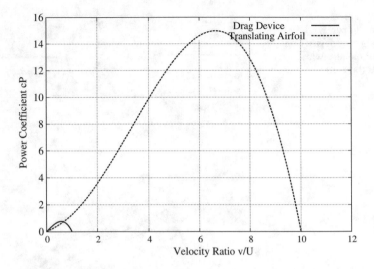

Fig. 2.18 Lifting device as compared to a pure drag device

$$P = \frac{2}{27}\rho A_r w^3 \cdot \frac{c_L^3}{c_D^2} , \qquad (2.15)$$

which gives

$$c_P = \frac{4}{27} L2D^3 c_D , \qquad (2.16)$$

a value which easily may exceed Betz' limit, because not the swept area but the wind area is used as reference To give numbers from a typical example [38] results from a demonstration plant with 4–8 m^2 wind area indicate values for c_L/c_D in the order 3, leaving much room for improvement.

2.9 Concluding Remarks

We finish this chapter by noting that still a myriad of other turbine types exist, some of which are based on very specific fluid mechanics principles or ideas. Even a pinwheel (toy) turbine was investigated experimentally [17]. The reader may consult the older literature [27, 36] for many other highly entertaining examples.

2.10 Problems

Problem 2.1 Derive Eq. 2.10 and find an expression for c as a function of c_{D+} and c_{D-}.

Problem 2.2 Estimate the increase of power for a DAWT with the following properties: $D_{Rotor} = 1$ m, $D_{exit} = 1.17$ m, and efficiency of diffuser $\eta_{diff} = 0.85$.

Problem 2.3 Determine if a counter-rotating turbine consists of one turbine or two, and give your reasons.

References

1. Ahlers U, Diehl M, Schmehl R (2014) Airborne wind energy. Springer, Berlin
2. Ashwill TD (1992) Measured data for the Sandia 34-meter vertical axis wind turbine, SAND91-228, Albuquerque, New Mexico, USA
3. Beurskens J (2014) History of wind energy, chapter 1 of: understanding wind power technology, Schaffarczyk AP (ed). Wiley Ltd, Chichester
4. Brecht B (2008) Life of Galileo. Penguin Classics, London. (Reprint)
5. van Bussel GJW (2007) The science of making more torque from wind: diffuser experiments and theory revisited. J Phys: Conf Ser 75:012010
6. Carne TG et al (1982) Finite element analysis and modal testing of a rotating wind turbine, SAND82-0345. Albuquerque, New Mexico, USA
7. Darrieus GJM (1931) Turbine having its Rotating shaft transverse to the flow field of the current, US Patent 1 835 018, 1931
8. Ferreira CS (2009) The near wake of the VAWT. TU Delft, The Netherlands PhD Thesis
9. FloWind Corporation (1996) Final project report: high-energy rotor development, test and evaluation, SAND96-2205, Albuquerque, New Mexico, USA
10. Herzog R, Schaffarczyk AP, Wacinski A, Zürchner O (2010) Performance and stability of a counter-rotating windmill using a planetary gearing: measurements and simulation. In: Proceedings of the EWEC 2010, Warsaw, Poland
11. Homicz GF (1991) Numerical simulation of VAWT stochastic aerodynamic loads produced by atmospheric turbulence: VAWT-SAL Code, SAND91-1124, Albuquerque, New Mexico, USA
12. International Electro-technical Commission (2011) IEC 61400-2 Ed. 3, Small wind turbines, Geneva, Switzerland (draft)
13. Kirke BK (1998) Evaluation of self-staring vertical axis wind turbines for stand-alone applications, PhD thesis, Griffith University, Gold Cost, Australia
14. Lilley GM, Rainbird WJ (1956) A priliminary report onthe design and performance of ducted windmills. Cranfield College of Aeronautics, Bedford
15. Loyd ML (1980) Crosswind kite power. J Energy 4(3):106–111
16. Mikkelsen R (2003) Actuator disk methods applied to wind turbines, PhD thesis, The Technical University of Denmark, Lyngby
17. Nemoto Y, Ushiyama I (2003) Experimental study of a pinwheel-type wind turbine. Wind Eng 27(2):227–235
18. Nossen P-O et al (2009) WIND POWER - the Danish Way. The Poul la Cour Foundation, Askov, Denmark
19. Oler JW et al (1983) Dynamic stall regulation of the Darrieus turbine, SAND82-7029, Albuquerque, New Mexico, USA
20. Paraschivoiu I (2002) Wind turbine design, with emphasis on Darrieus concept. Polytechnic International Press, Montreal, Canada,
21. Shedahl RE, Feltz LV (1980) Aerodynamic performance of a 5-meter-diameter Darrieus turbine with extruded aluminum NACA-0015 Blades, SAND80-0179, Albuquerque, New Mexico, USA
22. Shen WZ, Zakkam VAK, Sørensen JN, Appa K (2007) Analysis of counter-rotation wind turbines. J Phys: Conf Ser 75:012003

23. Spera D (ed) (2009) Wind turbine technology, 2nd edn. ASME Press, New York
24. Strickland JH (1975) The Darrieus turbine: a performance prediction model using multiple streamtubes, 2014 SAND75-0431, Albuquerque, New Mexico, USA
25. Templin RJ (1974) Aerodynamic performance theory for the NRC vertical-axis wind turbine, LTR-LA-160, NRC, Canada
26. Vollan A (1977) Aero elastic stability analysis of a vertical axis wind energy converter, EMSB-44/77, Dornier system, Immenstaad, Germany
27. de Vries O (1979) Fluid dynamic aspects of wind energy conversion, AGARDograph, No. 243, Neuilly sur Seine, France
28. Worstell MH (1978) Aerodynamic performance of the 17 meter diameter Darrieus turbine, SAND78-1737, Albuquerque, New Mexico, USA

In German

29. Bankwitz H et al (1975) Entwicklung einer Windkraftanlage mit vertikaler Achse (Phase I), Abschlußbericht zum Forschungsvorhaben ET-4135 A. Dornier system GmbH, Friedrichshafen, Germany
30. Binder G et al (1978) Entwicklung eines 5,5 m Ø-Windenenergiekonverters mit vertikaler Drehachse (Phase II), Abschlußbericht zum Forschungsvorhaben T-79-04. Dornier system GmbH, Friedrichshafen, Germany
31. Dekithsch A et al (1982) Entwicklung eines 5,5 m Durchmesser-Windenenergiekonverters mit vertikaler Drehachse (Phase III), Abschlußbericht zum Forschungsvorhaben T-82-086. Dornier system GmbH, Friedrichshafen, Germany
32. Eckert L, Seeßelberg C (1990) Analyse und Nachweis der 50 kW - Windnergieanlage (Typ Darrieus), MEB 55/90, internal report. Dornier GmbH, Immenstaad, Germany
33. Fritzsche A, Jürgensmeyer W, Obermayr E (1990) Auslegung einer Windenergieanlage mit senkrechter Drehachse im Leistungsbereich 350–500 kW, Abschlußbericht zum Forschungsvorhaben 0328958 A. Dornier GmbH, Friedrichshafen, Germany
34. Henseler H (1990) Eole-D MW Technologieprgrame Darrieus Windenergieanlagen Anpaßentwicklung, 2. Abschlußbericht zum Forschungsvorhaben 0328933 P, Dornier GmbH, Immenstaad, Germany
35. Meier H, Richter B (1988) Messungen an der Windkraftanlge DAWI 10 und Vergleich mit theoretischen Untersuchungen, Abschlußbericht WE-4/88 zum Forschungsvorhaben 03E–8384-A. Germanischer Lloyd, Hamburg, Germany
36. Molly J-P (1990) Windenergie - Theorie, Anwendung, Messung, 2nd edn. Verlag C.F. Müller Karsruhe, Germany
37. NN, Technische Anlage zum Angebot Nr. 3026-0-90, Lieferung und Montage einer 2,25 MW Darrieus-Windenergieanlage EOLE-D, Dornier GmbH, Friedrichshafen, Germany, 1990
38. Ranneberg M, Wölfle D, Bormann A, Rohde P, Breiopohl F, Basigkeit I (2018) Fast power curve and yield estimation of pumping airborne wind energy system. In: Schmehl R (ed) Airborne wind energy - advances in technology development and research. Springer, Singapore
39. Schaffarczyk AP (2007) Auslegung einer Kleinwindanlage mit Mantel, aerodynamischen Leistungsdaten, Optimierung des Diffsors, unveröffentlichte und vertrauliche Berichte Nr 49, 50 und 51
40. Savonius SJ (1930) Windrad mit zwei Hohlflügeln, deren Innenkanten einen zentralen Winddurchlaßspalt freigeben und sich übergreifen. Patentschrift Nr. 495:518
41. Soler A, Clever HG (1991) Bau, Aufstellung und Erprobung einer 50kW-Darrieus-Windkraftanlage, Abschlussbericht zum Forschungsvorhaben 0328726 P. Dornier GmbH / Flender Werft AG, Immenstaad und Lübeck, Germany in danish

In Danish

42. Clausen RS, Sønderby IB, Andkjær JA (2006) Eksperimentel og Numerisk Undersøgelse af en Gyro Turbine, Plyteknisk Midtvejsprojekt, Institut for Mekanik, Energi og Konstruktion, The Danish Technical University, Lyngby, Denmark

Chapter 3
Basic Fluid Mechanics

*Ich behaupte aber, daß in jeder besonderen Naturlehre nur so
viel eigentliche Wissenschaft angetroffen werden könne, als
darin Mathematik anzutreffen ist (Immanuel Kant, 1786) [32].
(However, I claim that in every special doctrine of nature there
can be only as much proper science as there is mathematics
therein. (Ref Stanford Encyclopedia of Philosophy)).*

3.1 Basic Properties of Air

Air regarded as an ideal gas may be described by its mass-density $\rho = dm/dV$.
To adjust its standard value of $\rho_0 = 1.225$ kg/m^3 to other temperatures (ϑ) and
elevations (H) we may use:

$$\rho(p, T) = \frac{p}{R_i \cdot T}, \tag{3.1}$$
$$T = 273.15 + \vartheta, \tag{3.2}$$
$$R_i = 287, \tag{3.3}$$
$$p(z) = p_0 \cdot e^{-z/z_{ref}}, \tag{3.4}$$
$$p_0 = 1015 \text{ hPa}, \tag{3.5}$$
$$z_{ref} = 8400 \text{ m}. \tag{3.6}$$

An other important parameter is the velocity of sound:

$$c = \sqrt{\kappa \cdot \frac{p}{\rho}} . \tag{3.7}$$

With $\kappa = 1.4$ (meaning that air consists mainly of dual-atom molecules) we have
under otherwise standard conditions ($p = p_0$) $c_0 = 340$ m/s. Usually flow is regarded
incompressible if $v/c \leq 1/3$. This means that the tip speed of a rotating wind turbine
blade has to be less than 113 m/s. At the time of this writing (December 2019), the

© Springer Nature Switzerland AG 2020
A. P. Schaffarczyk, *Introduction to Wind Turbine Aerodynamics*,
Green Energy and Technology, https://doi.org/10.1007/978-3-030-41028-5_3

largest rotor has a diameter of $D = 220$ m, which means RPM has to be limited to 9.8 to maintain incompressible airflow.

Temperature dependence of dynamic viscosity can be calculated by Sutherland's equation

$$\mu = \frac{C_1 \cdot T^{3/2}}{T + S} \,, \tag{3.8}$$

for air $C_1 = 1.5 \cdot 10^{-6}$ and $S = 110.4$ K. Or if pressure is included as well [29]

$$\mu = (1.82 \cdot (t/298.15))^{0.7}) + 1.6 \cdot 10^{-3} \cdot (P - 1) \cdot 10^{-7} \cdot (79 - (T - (273.15))) \cdot (P - 1)^2 \cdot 10^{-5} \,. \tag{3.9}$$

3.2 The Laws of Fluid Mechanics in Integral Form

3.2.1 Mass Conservation

Continuum mechanics [10] is formulated as a *local* theory for the velocity and (static) pressure field. Unless otherwise indicated, we use the *Eulerian* frame of reference where we assign these fields to points in an absolute (earth-fixed) frame of reference. To arrive at a global solution, we must integrate a set of partial differential equations (see Sect. 3.3) after which we know (in principle) all details of the flow field. In most engineering applications, forces on structures are important and may be derived from the pressure field by further integration.

Fortunately, this complicated affair may be circumvented by a direct discussion of the corresponding *conservation laws*, which are the basis of the local formulation.

We start with the equation of conservation of mass.

Let V be a finite Volume and ∂V its surface. Then a change of mass may only be possible if there is some flow across the boundary:

$$\dot{m} := \frac{d}{dt} \int_V \rho \cdot dV = - \int_{\partial V} \rho \mathbf{v} d\mathbf{A} \tag{3.10}$$

Here $\mathbf{v} = (\mathbf{u}, \mathbf{v}, \mathbf{w})$ is the velocity vector in Cartesian coordinates $\mathbf{r} = (x, y, z)$. The minus sign indicates that we count $d\mathbf{A} = \mathbf{n} dA$ positive *outwards*.

Referring to the example in Fig. 3.1 and assuming no perpendicular flow through the slip stream's surface, we clearly see that along the slipstream (1 = far upstream, 2 = disk, and 3 = far downstream):

$$v_1 \cdot A_1 = v_2 \cdot A_r = v_3 \cdot A_3 \tag{3.11}$$

must hold.

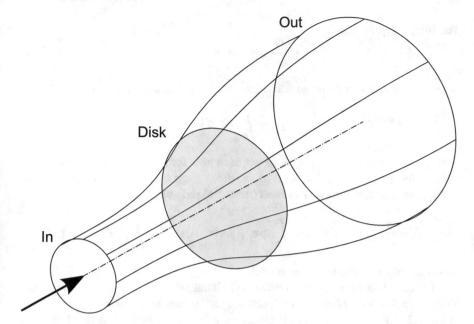

Fig. 3.1 Control volume and conservation of mass

3.2.2 Balance of Momentum

Newton's Second Law states that for a point-particle of mass m its momentum $\mathbf{p} := m \cdot \mathbf{v}$ may only by changed by the action of a force:

$$\mathbf{F} = \frac{d\mathbf{p}}{dt} .$$ (3.12)

Shifting to continuum mechanics, we first have to apply kinematics in an Eulerian frame of Ref. [10]: To get the acceleration

$$\mathbf{a}(t) := \frac{d}{dt}\mathbf{v}(x(t), y(t), z(t), t)$$ (3.13)

the *chain rule* of calculus has to be applied:

$$\mathbf{a}(t) = \frac{\partial \mathbf{u}}{\partial x}\dot{x} + \frac{\partial \mathbf{u}}{\partial y}\dot{y} + \frac{\partial \mathbf{u}}{\partial z}\dot{z} + \frac{\partial \mathbf{u}}{\partial t} .$$ (3.14)

in a more condensed manner we write

$$\mathbf{a}(t) = \partial_t\mathbf{v} + \mathbf{v} \cdot \nabla\mathbf{v} .$$ (3.15)

The formal expression

$$\frac{D}{Dt} := \partial_t + \mathbf{v} \cdot \nabla \tag{3.16}$$

is called *convective* or *material* derivative. Within a volume V the momentum is

$$\mathbf{p} := \int_V \rho \mathbf{v} \cdot dV \tag{3.17}$$

A slight complication arises because we have now forces acting on the volume (as volume forces $\mathbf{f} = d\mathbf{F}/dV$ as well as on the surface ∂V. If we (for the moment) restrict the analysis to normal stresses (pressure) only we have

$$\dot{\mathbf{p}} = \frac{d}{dt} \int_V \rho \mathbf{v} \cdot dV = - \int_{\partial V} p\mathbf{n} + \rho \mathbf{v}(\mathbf{v} \cdot \mathbf{n}) v dA + \int_V \rho \mathbf{f} dV . \tag{3.18}$$

if—as already remarked—$d\mathbf{A} = \mathbf{n} \cdot dA$.

The second summation term on the right-hand side of Eq. (3.18) may be called *momentum flux per unit area*. Its importance can be seen by the practical application in Fig. 3.1. With no change in total momentum $\dot{\mathbf{p}} = 0$ and no body forces $\mathbf{f} \equiv 0$, we see that if there is no pressure and no velocities outside the jacket:

$$\mathbf{T} + \dot{m}(-v_1 + v_3) . \tag{3.19}$$

Here

$$\mathbf{T} = - \int_{A_r} \Delta p \, d\mathbf{A} . \tag{3.20}$$

may be used to define the thrust **on** the disk (as a reaction force). It remains to be seen in Chap. 5 how v_1, v_2, and v_3 are related. However, from Eq. 3.19 it is clear that there **must** be forces applied to the disk if $v_3 < v_1$.

3.2.3   Conservation of Energy

Within a volume V the kinetic energy is

$$E_{kin} := \frac{1}{2} \int_V \rho \mathbf{v}^2 \cdot dV. \tag{3.21}$$

The energy content may only be changed by work done by pressure or body forces:

$$\dot{E}_{kin} = \frac{d}{dt} E_{kin} = - \int_{\partial V_t} p\mathbf{v}d\mathbf{A} + \int_{V_t} \rho \mathbf{v} \cdot \mathbf{f} dV . \tag{3.22}$$

A flow is called *isentropic* if there is a function h, such that

$$\nabla \cdot h = \frac{1}{\rho} \nabla p \;.$$ (3.23)

With this quantity the integral formulation of energy conservation may formulated as the famous **Bernoulli's Theorem**:

Theorem 3.1 *In stationary isentropic flow, the quantity H*

$$H = \frac{1}{2} \mathbf{v}^2 + h$$ (3.24)

remains constant along stream tubes.

Application of this energy equation to our *Actuator Disk* in Fig. 3.1 is then possible from $1 \rightarrow 2^-$ just in front of the disk and from immediately after the disk to a position from downstream $2^+ \rightarrow 3$. This is because we have seen that there must be a force and a pressure drop Δp at the disk (Fig. 3.2).

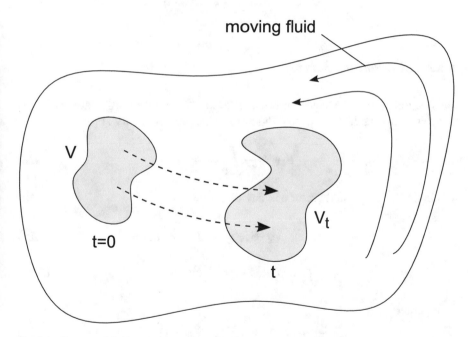

Fig. 3.2 Convected fluid and conservation of energy

3.3   Differential Equations of Fluid Flow

3.3.1   Continuity Equation in Differential Form

By applying the *Divergence Theorem* to a somewhat arbitrary vector field $\mathbf{A}$

$$\int_V \nabla \cdot \mathbf{A} \, dV = \int_{\partial V} \mathbf{A} \cdot d\mathbf{A} \tag{3.25}$$

to Eq. (3.10) we have the *continuity equation*

$$\frac{\partial \rho}{\partial t} + \nabla \cdot (\rho \mathbf{v}) = 0 \,. \tag{3.26}$$

If the density ρ is constant in space and time ($\rho(\mathbf{r}, t) = $ const.) this simplifies to a kinematic constraint only:

$$\nabla \cdot \mathbf{v} = \frac{\partial u}{\partial x} + \frac{\partial v}{\partial y} + \frac{\partial w}{\partial z} = 0 \,. \tag{3.27}$$

3.3.2   Momentum Balance

The integral momentum equation Eq. (3.18) may be put into a differential equation by again applying the Divergence Theorem to the pressure on ∂V. From

$$\mathbf{S}_{\partial \mathbf{v}} = -\int_{\partial V} p \, d\mathbf{A} \,. \tag{3.28}$$

we obtain by using a unit vector $\mathbf{e}$ in any direction:

$$\mathbf{e} \cdot \mathbf{S}_{\partial \mathbf{v}} = -\int_{\partial V} p \mathbf{e} \cdot d\mathbf{A} \tag{3.29}$$

$$= -\int_V \nabla \cdot (p\mathbf{e}) dV = \int_V (\nabla p) \mathbf{e} \, dV \,. \tag{3.30}$$

Together:

$$\frac{\partial \mathbf{v}}{\partial t} + (\mathbf{v} \cdot \nabla)\mathbf{v} = \frac{1}{\rho} (-\nabla p + \mathbf{f}) \,. \tag{3.31}$$

This is *Euler's equation* of flow.

Several remarks may be made as follows:

- the equation is **non**linear from the convective acceleration;

- it is first order in the velocity field, so only one boundary condition; at the solid wall must be applied
- in case of no volume forces it may be used to calculate the pressure field;
- as we have 1 (from mass conservation) + 3 (momentum balance) = 4 equations and 1 (pressure) + 3 (velocity) unknown fields the problem is at least not indeterminate.

3.3.3 Differential Energy Equation

For incompressible fluids, or to be more exact, in flow conditions where the compressibility may be neglected ($v/c \leq 1/3$), mass and momentum balance provides a sufficient number of equations. Therefore, the differential energy equation is trivial:

$$\frac{D\rho}{Dt} = 0 . \tag{3.32}$$

This situation changes only if $\rho = \rho(\mathbf{r}, t)$ becomes variable. In addition thermodynamics enters the picture, as air then must be regarded at least as an ideal gas and $\rho = \rho(T, p)$ as mentioned earlier.

3.4 Viscosity and Navier–Stokes Equations

The early description of fluid motion staring with D. Bernoulli's *Hydrodynamics* [13] was hopelessly wrong in describing practical applications, as shown by the aforementioned Paradox of D'Alembert Theorem 2.1. One important step toward a more *realistic* description, in the form of accurate numerical prediction, becomes possible with the inclusion of internal friction or viscosity. Under one-dimensional flow conditions, it simply states there must be shear stress to maintain cross-gradients:

$$\tau := \eta \cdot \frac{du}{dy} . \tag{3.33}$$

See Fig. 3.3. A fluid fulfilling this property is called a *Newtonian* fluid. A generalization to three dimensions is straightforward. Instead of (x, y, z) for the Cartesian components, we will include indices for the coordinates (x_1, x_2, x_3). Latin indices (i, j, k, ...) then obey $1 \leq i, j, k \leq 3$ and $(i, j) \in \mathbb{N}$. The Cauchy stress tensor

$$\sigma = (\sigma_{ij}) = \begin{pmatrix} \sigma_{11} & \sigma_{12} & \sigma_{13} \\ \sigma_{21} & \sigma_{22} & \sigma_{23} \\ \sigma_{31} & \sigma_{32} & \sigma_{33} \end{pmatrix} \tag{3.34}$$

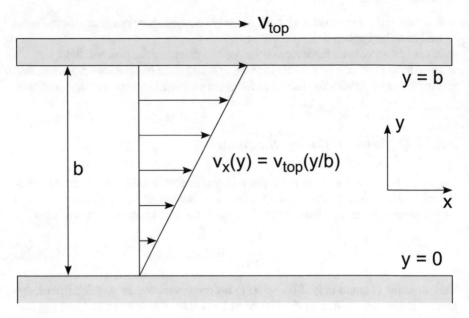

Fig. 3.3 On the definition of viscosity

simply gives forces (index 1) in direction i with reference to directions (index 2) j. With $\mathbf{v} = (u, v, w) \rightarrow (v_1, v_2, v_3)$ we define a deformation tensor[1]

$$\mathbf{D} = (D_{ij}) = \frac{1}{2} \left(\frac{\partial v_i}{\partial x_j} + \frac{\partial v_j}{\partial x_i} \right) \tag{3.35}$$

Now the 3D generalization of Eq. (3.33) is

$$\sigma = 2\eta \cdot \mathbf{D} . \tag{3.36}$$

Now after some algebra we finally arrive at the *Navier–Stokes Equations* (NSE).

$$\frac{\partial \mathbf{v}}{\partial t} + (\mathbf{v} \cdot \nabla)\mathbf{v} = \frac{1}{\rho} (-\nabla p + \mathbf{f}) + \nu \Delta \mathbf{v} . \tag{3.37}$$

Here $\Delta = \nabla^2$ is the Laplacian Operator which acts on each component and

$$\nu = \frac{\eta}{\rho}, \tag{3.38}$$

---

[1]A tensor may be represented by a n × n matrix. However, like scalars and vectors it is defined by its transformation rules under change of coordinates.

the *kinematic viscosity*. By now it generally believed that all types of flow—as long as the fluid is Newtonian—may be described by Eq. 3.37. We will come back to this discussion when talking about *turbulence* in Sect. 3.8. Again we close with some remarks:

- NSE are **non**linear.
- they are of **second** order, so two boundary conditions at the solid wall must be provided. This defines the so-called *slip condition*, which means that normal (as in the case of Eulerian flow) as well as tangential components of velocity have to vanish at solid walls.
- the limit $\nu \to 0$ is highly nontrivial as it changes the order of the differential equation. As this limit is at the heart of *Boundary Layer Theory*, we will discuss it in more detail in Sect. 3.7.

3.5 Potential Flow

3.5.1 General 3D Potential Flow

Another very important quantity is *vorticity*. It is defined by

$$\omega := \nabla \times \mathbf{v} \, . \tag{3.39}$$

and is closely related to *circulation*:

$$\Gamma_{C_t} := \oint_{C_t} \mathbf{v} \cdot d\mathbf{r} \, . \tag{3.40}$$

which may be seen using *Stokes Theorem* for a vector field $\mathbf{B}$ on a surface A with boundary $\partial A = C_t$:

$$\int_{\partial A} (\nabla \times \mathbf{B}) \cdot d\mathbf{A} = \oint_A \mathbf{B} \cdot d\mathbf{r} \, . \tag{3.41}$$

This establishes the close relationship between both quantities. A flow is called *irrotational* if $\nabla \times \mathbf{v} = 0$ everywhere. If we assume there exists a scalar function φ with

$$\mathbf{v} := \nabla \Phi \, , \tag{3.42}$$

we call Φ a *velocity potential*. With some weak assumption on the differentiability on Φ we get from $\nabla \times \nabla \varphi = 0$ directly $\nabla \times \mathbf{v} = 0$, which means that such a flow is *irrotational*. This holds only if the region in which the flow is irrotational is simply connected, which means in simple terms that the region does not have holes. Note that even one point may be regarded as a hole (Fig. 3.4).

Fig. 3.4 Example for a
multi-connected region

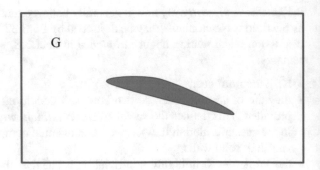

In such cases

Theorem 3.2 *In isentropic flow, the circulation,* Γ_{C_i}, *Eq. (3.40) is a conserved quantity—constant in time holds [2, 10, 48]. Potential theory is especially useful and beautiful in two dimensions. Therefore a subsection on this topic is appropriate. Because of*

$$\nabla \cdot (\nabla \times \mathbf{A}) \equiv 0 \qquad (3.43)$$

The Continuity Equation is fulfilled identically if we set

$$\mathbf{v} \equiv \nabla \times \mathbf{A} . \qquad (3.44)$$

$\mathbf{A}$ is then called *vector potential*. We will come back to this useful quantity in Sect. 6.6. It has to be noted that 3D potential theoretical calculations were very popular [34] in the early days of *computational fluid mechanics*. As they predict forces resulting only from vortex-induced flow (see Sect. 3.6) they may serve only as to give an impression of the general flow field. Nevertheless, it will be seen in the context of *boundary layer flow* (see Sect. 3.7) that—in case of non-*separated* flow—except for a thin layer, most of the flow may be regarded as potential flow.

3.5.2   2D Potential Attached Flow

In this subject, we will explain how useful and easy potential flow methods are in two dimensions. To obey the 2D continuity equation (from Eq. 3.27 by omitting z and w): we have

$$\frac{\partial u}{\partial x} + \frac{\partial v}{\partial y} = 0 , \qquad (3.45)$$

a stream function

$$\psi = \psi(x, y) \qquad (3.46)$$

is introduced. It is easily seen, that after canceling terms the only remaining component of the 3D vector-potential, Eq. 3.44:

$$u = \frac{\partial \psi}{\partial y} \tag{3.47}$$

$$v = -\frac{\partial \psi}{\partial x} \; . \tag{3.48}$$

Now ψ as well as ϕ (the vector potential) obey

$$\Delta \Psi = \Delta \Phi = 0 \; . \tag{3.49}$$

Now introducing *complex variables* $z := x + i \cdot y \in \mathbb{C}$: together with complex
velocity $\mathbb{C} \ni w = u - i \cdot v$ and *complex potential* $F(z) := \Phi + i \cdot \Psi$. An important result from complex calculus is, that (by the action of *Cauchy–Riemann's*
Theorem [34].

Theorem 3.3 *Every complex function* $A(z) = u(x, z) + i \cdot v(x, y)$ *which is homo-
morphic (complex differentiable) broken down to real and imaginary parts satisfies
the potential equation*

$$\Delta u = \Delta v = 0 \; . \tag{3.50}$$

Now let $\mathbb{C} \to \mathbb{C}$ with $\zeta \to z f(\zeta)$ via

$$\zeta \to := z + \frac{c^2}{z} \tag{3.51}$$

the *Joukovski Transformation* (see Problem 3.4 for details) which maps a circle
$z(\phi) = z_m + R \cdot e^{i \cdot \phi}, 0 \le \phi \le 2\pi$ to the profile in Fig. 3.5.

The linearity of the potential equation allows us to superimpose various solutions.
For a cylinder of radius in uniform flow with $w = u + i \cdot 0$ we have

$$F(z) = u \cdot z + \frac{R^2 \cdot u}{z} \tag{3.52}$$

which can be transformed to a corresponding pressure distribution on the profile, see
Fig. 3.6. As Eq. 3.51 is holomorphic except for $\mathbb{C} \ni z = 0$ its inversion is as well.
Therefore a one-to-one relationship between pressure distribution and profile shape
exists, which may be extended to a design tool for aerodynamic profiles [14, 17].

It was not until the advent and distribution of these kinds of simple PC-codes that
special wind turbine profiles, see Sect. 9.2, were developed. Now adding a potential
vortex to Eq. (3.52):

$$F(z) = u \cdot z + \frac{R^2 \cdot u}{z} - i \cdot \Gamma \; log(z) \; , \tag{3.53}$$

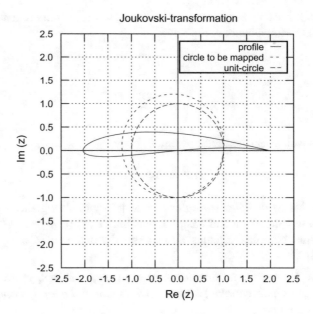

Fig. 3.5 Generation of a Joukovski profile by conformal mapping

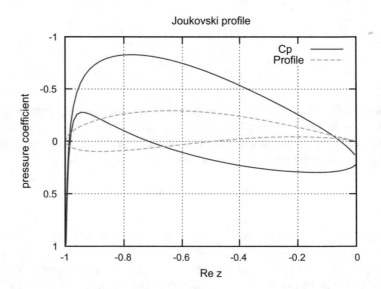

Fig. 3.6 Pressure distribution of a simple Joukovski profile

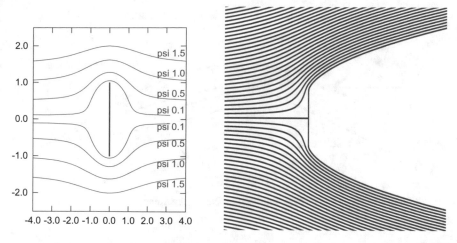

Fig. 3.7 Flow normal to a plate (red). Attached (left $c_D = 0$) and separated (right, $c_D = 0.88$) potential-theoretic flow

we are able to generate *Lift*. From Blasius'[2] theorem [10] using a complex force $\mathscr{F} = F_x + i \cdot F_y$ and $\overline{\mathscr{F}} = F_x - i \cdot F_y$:

Theorem 3.4 *In exterior flow:*

$$\overline{\mathscr{F}} = -\frac{i\rho}{2} \int_{\partial\mathscr{B}} w^2 \, dz \, . \tag{3.54}$$

From this very general theorem, we obtain if we include circulation (positive in z-direction):

Theorem 3.5 *If there is circulation $\Gamma = \oint \mathbf{v} \cdot d\mathbf{r}$ around $\mathscr{B}$ with inflow velocity $\mathbf{v}_\infty$, then a force (positive in y-direction)*

$$L = \rho \cdot \mathbf{v}_\infty \cdot \Gamma \tag{3.55}$$

emerges (Fig. 3.7).

The amount of circulation may be determined by the following corollary:

Corollary 3.1 *The flow around an airfoil has to be of finite (but not continuous) velocities [59].*

---

[2]Heinrich Blasius, ⋆ 08.09.1883, † 04.24.1970, was one of *L. Prandtl's* first Ph.D. students. His work on forces and *Boundary Layer* flat plate theory is discussed in almost every textbook of fluid mechanics even today. Less known is that he taught over 50 years (1912–1970) at *Ingenieuerschule* (Polytechnic—now University of Applied Sciences) Hamburg.

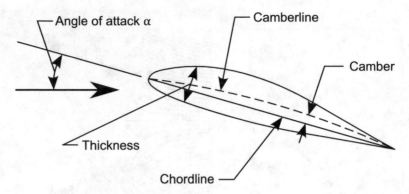

Fig. 3.8 Angle of attack and other geometrical properties regarding an airfoil section

This more or less mathematical statement[3] gave the first explanation of lift which for a long time was a mystery. Even today many incorrect explanations exist. For an easy-to-read version see [4], [7] of Chap. 1. An extension of these ideas gave rise to *Thin Airfoil Theory* , originated by *Max Munk* [50], *Hermann Glauert* [25], and others (see Sect. 3.6.2).

One of the fundamental results is with c_L being the *lift-coefficient* and α the *angle-of-attack*, see Fig. 3.8.

$$c_L = 2\pi \cdot \alpha . \tag{3.56}$$

One important ingredient is the *Kutta-Condition* which fixes the otherwise indeterminate amount of circulation by a simple mathematical condition that all velocities have to be bounded.

3.5.3   2D Potential Flow Behind a Semi-infinite Set of Lamina

Prandtl and Betz, see [36] in Chap. 6, [2] in Chap. 5 and Fig. 5.10 used the complex potential

$$F(z) = -v \, \frac{d}{\pi} \, arccos(exp(\pi z/d)) \tag{3.57}$$

with v the inflow and d the spacing of the plates.

---

[3]It may be of interest to note that here the essential argument is to avoid singularities in terms of infinite velocities, whereas in other applications this does not apply, for example, the infinite pressure at an inclined flat plate.

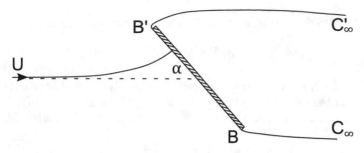

Fig. 3.9 Flow pattern of Helmholtz–Kirchhoff flow on an inclined flat plate, adapted from [48]

3.5.4 2D Separated Flow

As the result of further progress in complex calculus, Kirchhoff and Helmholtz [48] were also able to investigate separated flow. One success was the approximate formulation of drag on a inclined flat plate:

$$c_D = \frac{2\pi \cdot sin(\alpha)}{4 + \pi \cdot sin(\alpha)}, \tag{3.58}$$

which at $90°$ gives $c_{D,90} = 0,88$. The experimental value varies according to Reynolds number and an aspect ratio between 1.2 and 2.0. An [41, 53] application of this model pertaining to *dynamic stall* is available (Fig. 3.9).

3.6 Formulation of Fluid Mechanics in Terms of Vorticity and Vortices

The pressure may be eliminated from the body-force-free NSE, Eq. (3.37) by taking the *curl* of **v**

$$\frac{D\omega}{Dt} = \omega \cdot \mathbf{v} + \nu \Delta \omega . \tag{3.59}$$

As an outcome of pressure there may be computed only from a *Poisson's equation* which is a nonhomogeneous Laplace equation (Here we use Einstein's summation convention, to sum over all dual-indexed variables: $u_i \cdot u_i := \sum_{i=1}^{N} u_i \cdot u_i.$):

$$\Delta p = -\rho \frac{\partial^2 u_i u_j}{\partial x_i \partial x_j} \equiv S(\mathbf{r}) \tag{3.60}$$

Formally this equation then may be solved by introduction of *Green's function*:

$$p(\mathbf{r}) = p^{harmonic}(\mathbf{r}) + p(\mathbf{r}) + \frac{\rho}{4\pi} \int_{\mathbb{R}^3} \frac{S(\mathbf{r})}{|\mathbf{r} - \mathbf{r}'|} \, d\mathbf{r}' , \tag{3.61}$$

where $p^{harmonic}(\mathbf{r})$ is a solution of the homogenous pressure equation:

$$\Delta p^{harmonic}(\mathbf{r}) = 0 . \qquad (3.62)$$

Obviously the situation becomes much easier if vorticity is concentrated. If so, these compact distributions then may be regarded as the *sinews and muscles of fluid motion* [59]. The simplest case is a so-called *point-vortex* (2D) and *line-vortex* (3D). From Eq. (3.42) we get

$$\omega = \nabla \times \mathbf{v} = -\nabla^2 \mathbf{A} \qquad (3.63)$$

if $\nabla \cdot \mathbf{A} = 0$. Again by using *Green's functions*:

$$\mathbf{A} = \frac{1}{4\pi} \int_{\mathbb{R}^3} \frac{\omega}{|\mathbf{r} - \mathbf{r}'|} \, dV \qquad (3.64)$$

which is called *Biot–Savart Law*. Inserting it into the velocity equation we get

$$\mathbf{v} = \frac{1}{4\pi} \int_{\mathbb{R}^3} \nabla \times \frac{\omega}{|\mathbf{r} - \mathbf{r}'|} \, dV . \qquad (3.65)$$

We apply this now to a specific situation in which ω is only nonzero along a line, for example, the z-axis in a 3D Cartesian system. Performing the integration (see Problem 3.2) and referring to Fig. 3.10 we arrive at (Figs. 3.11 and 3.12):

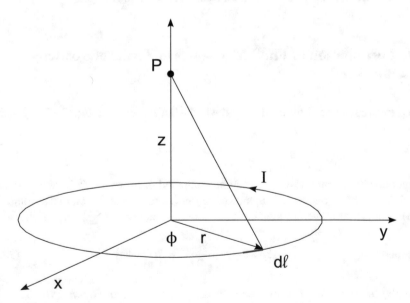

Fig. 3.10 Biot–Savart law

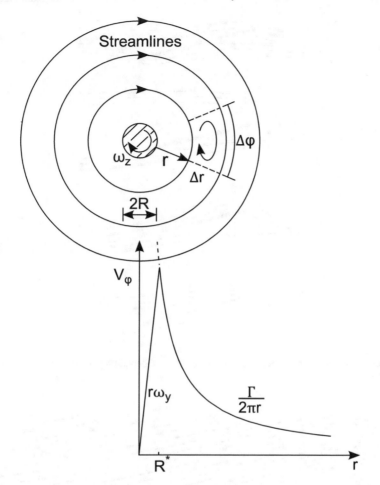

Fig. 3.11 Point(2D) or Line(3D) vortex

$$v_r = 0, \tag{3.66}$$
$$v_\phi = \frac{\Gamma}{2\pi r}, \tag{3.67}$$

with Γ from Eqs. (3.39) to (3.41).

3.6.1   Flow and Forces

See Fig. 3.13 and Table 3.1.

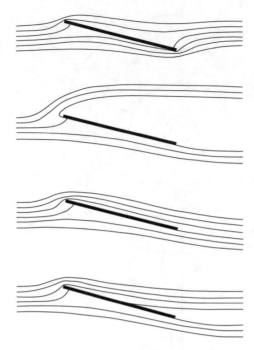

Fig. 3.12 Origin of forces induced by an inclined flat plate depending on the flow pattern. Adapted from Jones [31]

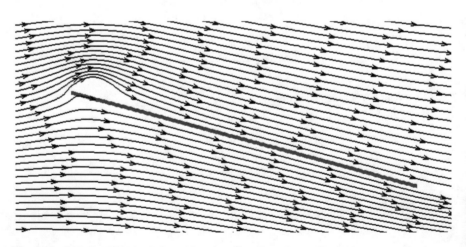

Fig. 3.13 CFD (RANS) simulation of flat plate flow, according to Fig. 3.12

Table 3.1 Numerical values

Type	c_L	c_D	Remark
Potential theory	0.00	0.00	
Helmholtz–Kirchhoff	0.33	0.09	
Kutta–Joukovski	1.63	0.00	
BL-theory (1)	1.53	0.02	NACA0015 XFoil
BL-theory (2)	0.70	0.25	NACA0005 XFoil
RANS laminar	0.92	0.25	FLUENT
RANS turbulent	1.03	0.30	FLUENT $k - \varepsilon$
Measurement	0.81	0.22	Föppl [21]

3.6.2 Thin Airfoil Theory

With [25] we have a 2D section of chord c where the circulation is distributed along a so-called *skeleton line* $y(x)$, $0 \le x \le c$ of

$$\Gamma = \int_0^c \gamma(x)\, dx \ . \tag{3.68}$$

According to Biot–Savart's Law Eq. (3.65), we have induced velocities:

$$v(x') = \int_0^c \frac{\gamma(x)\, dx}{2\pi \ (x - x')} \ . \tag{3.69}$$

If the inflow velocity is designated by V then by noting that no normal flow through the airfoil is possible, the important constraint is

$$\alpha + \frac{v}{V} = \frac{dy(x)}{dx} \ . \tag{3.70}$$

Glauert [25] solved this equation by expanding $\gamma(x)$ in terms of a *Fourier Series*:

$$\gamma(x) = 2V \left(A_0 \cdot \cot\left(\frac{\theta}{2} + \sum_{n=1}^{\infty} A_n \sin(n\theta)\right) \right) \tag{3.71}$$

$$x(\theta) = \frac{1}{2}c(1 - \cos(\theta))\ 0 \le \theta \le \pi \ . \tag{3.72}$$

He found:

$$c_L = 2\pi \left(A_0 + \frac{1}{2}A_1 \right) \tag{3.73}$$

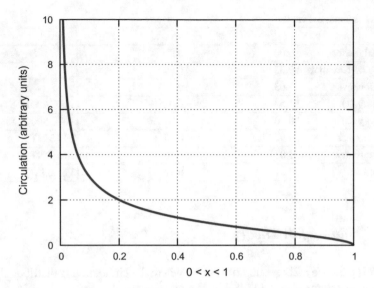

Fig. 3.14 Circulation distribution for an inclined flat plate

$$A_0 = \alpha - \frac{1}{\pi} \int_0^\pi \frac{dy(x)}{dx} d\theta \text{ and} \tag{3.74}$$

$$A_n = \frac{2}{\pi} \int_0^\pi \frac{dy(x)}{dx} \cos(n\theta) d\theta \ . \tag{3.75}$$

For a flat plate $A_0 = 1$ and $A_1 \approx 0$. The corresponding circulation distribution $\gamma(x)$ is shown in Fig. 3.14. It has a square-root singularity for $x \to 0$. The Kutta(–Joukovski) condition is implemented via $\gamma(x = 1) = 0$, as can be seen in Fig. 3.14.

3.6.3   Viscous Thin Airfoil Theory

TAT is based on inviscid models of fluid flows (only density as a material enters) and as a consequence, circulation—among others—is a conserved quantity, i.e., it can neither be created nor destroyed. Therefore, more sophisticated models (and equations) must be included if the emergence of lift is to be explained. It is well known that the Navier–Stokes equations provides this basis, adding a second material parameter, viscosity. In a series of journal and technical papers Yates [65] among others was able with the help of an Oseen-type approximation (in fact a linearization) using these Navier–Stokes equation to

1. derive the Kutta condition and
2. to give asymptotic corrections to the lift-curve slope in terms of inverse Reynolds number.

This is somewhat surprising as Oseen flow [2], Sect. 4.10, is generally assumed to be valid in low-Re (RN < 1, creeping) flow only, whereas in high-Re flow (RN > 10^5) boundary layer theory Sect. 3.7 should be more appropriate.

3.6.3.1 Finite Thickness

A lot of authors including Abbot and von Doenhoeff [1] tried to improve TAT by investigating the influence of thickness to lift curve slope which typically results in equations like [1],

$$c_L = 2\pi \left(1 + \tau\right) , \tag{3.76}$$

$$\tau = \frac{\varepsilon}{a} = \frac{4\sqrt{3}}{9} \cdot \frac{t}{c} . \tag{3.77}$$

Yates [65] further gave Reynolds number corrections

$$c_L = 2\pi \left(1 + \tau\right) \cdot \left(1 - \frac{4}{\log\left(64 Re\right) + \gamma_E}\right) \tag{3.78}$$

$\gamma_E = 0.57722$ being Euler's constant which shows a decrease of about 10% at t/c = 0.3. from RN effects which—at least—is partly compensated by the first (thickness) term. No negative lift curve slope values are likely. McLean [7, pp. 313/314] in Chap. 1 gives further details.

Not included in all these discussions is the influence of the flow-state of the boundary layer, whether it is laminar or turbulent. In our discussions, we assume that lift (in the linear part) is not influenced as strong as drag. It is well known that drag can be much higher when most parts of the boundary layer are turbulent.

Another important phenomena, flow separation, the starting point defined by

$$c_f = \mu \cdot \frac{dv_t}{dn} \leq 0 , \tag{3.79}$$

is closely related to limiting c_L to values around 1.5 (of course, with remarkable exceptions).

3.6.4 Vortex Sheets

A special kind of 2D extended vortical objects are called *vortex sheets*. As can be shown [55, 59] this is more or less equivalent to a discontinuity in tangential velocity only. These sheets may be created, for example, by the sudden movement of a *Kaffelöffel* (coffee spoon) [35]. A major application is Prandtl's *lifting line theory*. We will come back to this point later in Sect. 6.5.

3.6.5 Vorticity in Inviscid Flow

As will be presented in more detail in Chap. 6, Theorem 6.1 (*Helmholtz' Theorem*), circulation is a conserved quantity in inviscid flow. In addition, along streamlines energy is also conserved when expressed in terms of enthalpy H, Eq. (3.23). H varies when changing streamlines according to *Crocco's equation*

$$\mathbf{u} \times \omega = \nabla \cdot H .\tag{3.80}$$

3.7 Boundary Layer Theory

3.7.1 The Concept of a Boundary Layer

As was remarked earlier, NSE and the inviscid Euler Equation are not connected by an analytical limit $\nu \to 0$, because the order of both differential equations is not the same. However, a very useful approximation is possible as will be seen in the following. We start with one of the simplest cases, the flat plate, which is described by uniform flow $u(y) = u_\infty = const, x < 0$ and $u(x, y = 0) = v(x, y = 0) = 0, x > 0$. We follow closely, but not completely, the path presented in [61]. We start with the stationary 2D Cartesian NSE $\mathbf{v} = (u,v)$ and set $\rho = 1$. Then we have the following set of equations:

$$\frac{\partial u}{\partial x} + \frac{\partial u}{\partial y} = 0,\tag{3.81}$$

$$u\frac{\partial u}{\partial x} + u\frac{\partial u}{\partial y} = -\frac{\partial}{\partial x}p + \nu\Delta u,\tag{3.82}$$

$$u\frac{\partial v}{\partial x} + v\frac{\partial v}{\partial y} = -\frac{\partial}{\partial y}p + \nu\Delta v.\tag{3.83}$$

As usual in 2D we obey continuity by introducing a stream function ψ:

$$u = \frac{\partial \psi}{\partial y},\tag{3.84}$$

$$v = -\frac{\partial \psi}{\partial x} .\tag{3.85}$$

We try a *separation approach* via

$$\psi := \sqrt{2\nu x u_\infty} \cdot f(y) .\tag{3.86}$$

In addition we use a *similarity approach* for the vertical coordinate (η now not to be confused with the earlier defined dynamic viscosity)

$$\eta := y\sqrt{\frac{u_\infty}{2vx}} . \tag{3.87}$$

Therefore:

$$u = x \cdot f', \quad v = -f, . \tag{3.88}$$

Inserting into Eqs. (3.82) and (3.83), we may drop the pressure to arrive at *Blasius' Equation*:

$$f''' + ff'' = 0 . \tag{3.89}$$

At the wall $y = 0$ which means $f(0) = f'(0) = 0$ to obey the *no-slip* condition $u = v = 0$. We now have a third-order differential equation, so we have to add one boundary condition. We demand asymptotically for $y \to \infty$ or equivalently $\eta \to \infty$

$$f'(\eta) = 1 \text{ as } \eta \to \infty . \tag{3.90}$$

Blasius originally [7] solved Eq. (3.89) by a power-series approach. By now the mathematical understanding is much more complete [5], and very easy numerical schemes exist. The boundary condition (3.90) may be shifted to $f''(0) = const$ and looking for a constant $const$ which leads to $lim_{\eta \to \infty} f(\eta) = 1$. As may be seen from Fig. 3.15 at $\eta = 3.8$ (5) we have $f'(\eta) = 0.996$ (0.99994) so that we may define a *boundary layer thickness* by

$$\delta_{99\%} := 3.5 \cdot \sqrt{2vxu_\infty} . \tag{3.91}$$

Two other variables, displacement and momentum thickness, are also used:

$$\delta^\star := \int_0^\infty \left(1 - \frac{u}{u_\infty}\right) dy, \tag{3.92}$$

$$\theta := \int_0^\infty \frac{u}{u_\infty} \cdot \left(1 - \frac{u}{u_\infty}\right) dy . \tag{3.93}$$

Now we are able to calculate shear stresses on the wall:

$$c_f := \frac{2\tau_w}{\rho u_\infty^2} = \frac{0.664}{\sqrt{Re_x}} = \frac{\theta}{x} . \tag{3.94}$$

Rather accidentally we introduced a very important number the *Reynolds number*

$$Re_x := \frac{u_\infty \cdot x}{v} . \tag{3.95}$$

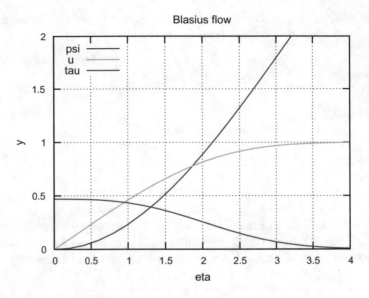

Fig. 3.15 Boundary layer at a flat plate—Blasius' Solution

With typical numbers from wind turbines (x = 1 ... 10 m, u = 10 ... 50 m/s, $\nu =$ 1.5 · 10^{-5} m^2/s) we have $Re_x \sim 0.6...35 \cdot 10^6$, therefore BL-thicknesses on the order of cm. Figure 3.16 shows a comparison of measurement of friction with the theory developed so far. Two things are remarkable:

- The calculated c_F value fit very well in the range $Re < 10^5$ but starts to deviate at values less than 10^2
- For $Re > 10^5$ c_F is much larger, but a theory is also available.

3.7.2   Boundary Layers with Pressure Gradient

As has been seen for the case of a simple Joukovski airfoil, pressure is not constant on these important 2D sections. Therefore it is very useful to look for a simple (potential) flow which also may be calculated by BL-methods. This kind of flow is so-called wedge flow or *Falkner–Skan Flow*, see Fig. 3.17. The pressure may be characterized by a dimensionless number Λ and the profiles within the boundary layer are shown in Fig. 3.18. As may be expected, beginning separated flow ($\Lambda \approx 12$) exhibits a profile with a vertical tangent.

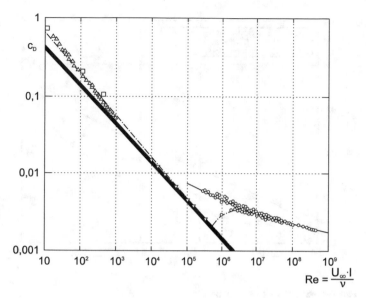

Fig. 3.16 Comparison of measurement and boundary layer theory

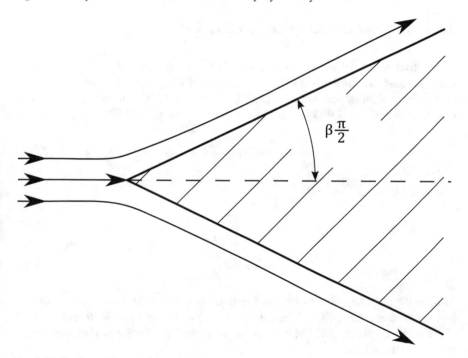

Fig. 3.17 Faulkner–Skan flow

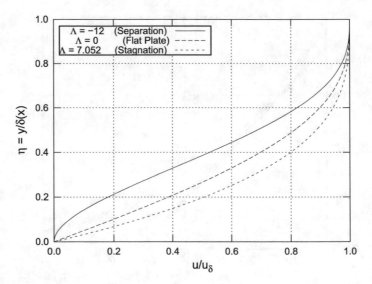

Fig. 3.18 Faulkner–Skan flow: velocity profiles

3.7.3   Integral Boundary Layer Equations

Differential equations are useful if all details of the flow are sought. In most cases, however, only the shear stress at the wall is needed. Considerable simplification can be achieved when integrating the BL-DEQ, Eq. (3.82) [34]. After some algebra one arrives at an ordinary DEQ for the momentum thickness Eq. (3.93)

$$\frac{d\theta}{dx} + (H + 2)\frac{\theta}{u_e}\frac{du_e}{dx} = \frac{C_f}{2} . \tag{3.96}$$

And using Eqs. 3.92 and 3.93:

$$C_f := \frac{\tau_w}{\frac{1}{2}\rho u_e^2} , \tag{3.97}$$

$$H := \frac{\delta^\star}{\theta} . \tag{3.98}$$

u_e is the edge velocity which also predicts the pressure at the BL-edge. As may be readily seen, this DEQ contains three unknown (functions): $\delta^\star$, θ and the wall shear stress τ_w via the friction coefficient. To solve for velocity profile, families of polynomials are used:

$$\frac{u}{u_e} = a_1\eta + a_2\eta^2 + a_3\eta^3 + a_4\eta^4 + \cdots 0 \le \eta \le 1 . \tag{3.99}$$

Applying the boundary conditions for the flat plate (Blasius) case, this reduces to

$$\frac{u}{u_e} = 2\eta - 2\eta^3 + \eta^4 \ . \tag{3.100}$$

Integration of Eq. (3.96) can be performed easily with the result for the friction coefficient, e.g.,

$$C_f \cdot \sqrt{Re_x} = 0.6854 \ , \tag{3.101}$$

instead of

$$C_f \cdot \sqrt{Re_x} = 0.664 \ , \tag{3.102}$$

for the exact Blasius solution. This method has been successfully implemented into XFoil (open-source, [14]) and Rfoil [56], the latter being enhanced for rotational effects which may become important at the root sections of WT-blades. Figure 3.19 shows as an example of the results of lift and drag for the Joukovski airfoil from Fig. 3.5.

Where do boundary layers occur in wind turbine aerodynamics? Two places are to be noted first and foremost: the lower atmosphere may be regarded as a boundary layer (with a thickness of several hundred meters, however) and when investigating flow over a wind turbine blade in more detail. Unfortunately, the latter BLs are of a very different kind—they are *turbulent* , which will be explained in Sect. 3.9.

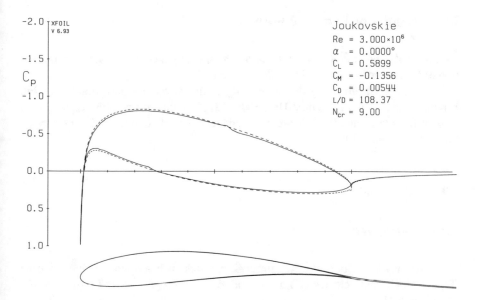

Fig. 3.19 Lift and drag data from integral BL-theory (XFoil) for a Joukovski airfoil

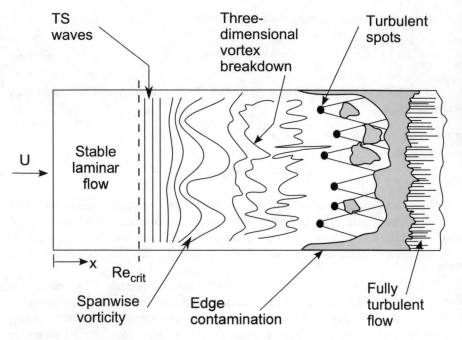

Fig. 3.20 Transition form instable laminar to fully turbulent flow

3.8 Stability of Laminar Flow

Instability of laminar flow is one precondition for emergence of a very different other type of flow, turbulent flow (Figs. 3.20, 3.21 and 3.22).

It is somewhat surprising that a stability analysis was not performed for wind turbine blade flow until recently. Hernandez [27] modified the stability equations to account for rotational effects. More about his results applied to transition prediction will be found in Sect. 8.7.

3.9 Turbulence

3.9.1 Introduction

The NSE may be written in a nondimensional version if reference values of time (T), length (l), and velocity (U) are introduced: $x \rightarrow x/l, t \rightarrow t/T, (u, v, w) \rightarrow u/U, v/u, w/U$:

$$\frac{\partial \mathbf{v}}{\partial t} + (\mathbf{v} \cdot \nabla)\mathbf{v} = (-\nabla p + \mathbf{f}) + \frac{1}{Re}\Delta \mathbf{v} , \qquad (3.103)$$

with $Re := \frac{U \cdot L}{\nu}$ and pressure in nondimensionalized form.

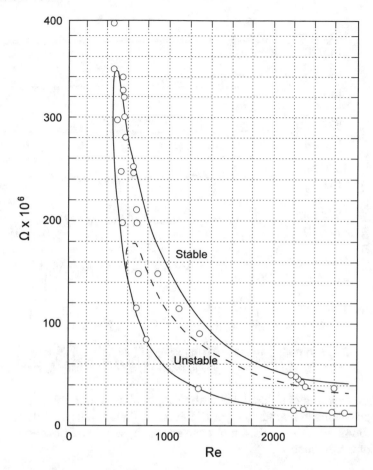

Fig. 3.21 Results from linear stability theory (lines) compared to measurements (dots) for flat plate flow, $Re^{cr}_{\Theta} \sim 520$

It was recognized very early by *Leonardo da Vinci* that a special (highly disordered—or as he called it *hairy*) state of fluid flow exists and he actually introduced the term (see Fig. 3.23) more than 500 years ago.

But it was not before the end of the nineteenth century that a scientific fluid-dynamical investigation was undertaken by *Osborne Reynolds*. He injected dye into a pipe of circular cross section and found that in case of high flow velocity, the shape and structure changed into an irregular and rapidly changing pattern (see Fig. 3.24). In addition to that he introduced *Reynolds averaging* so that a special kind of NSE is now abbreviated as *RANS* (Reynolds Averaged Navier–Stokes Equation).

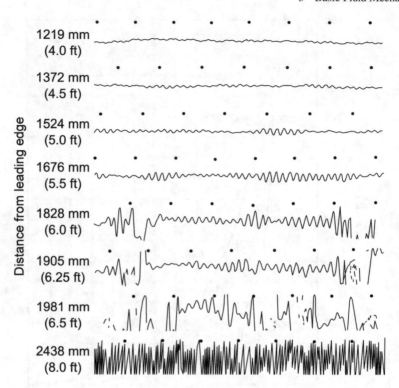

Oscillograms of u-fluctuations showing laminar boundary-layer oscillations in boundary layer of flat plate. Distance from surface, 0.5842 mm; U_∞= 24384 mm per second; time interval between dots, 1/30 second.

Fig. 3.22 First measurements of Tollmien–Schlichting waves in a flat-plate boundary layer, data from [62]

3.9.2 *Mathematical Theory of Turbulence*

3.9.2.1 Existence and Properties of Solutions

Many people believe that such investigations are of minimal importance. This is not at all true. A sound mathematical model serves in all known cases as a rich source for improving computation efforts. As explained in [19, 20, 24, 44] the mathematical theory of the properties of NSE is of utmost importance, because there is a still unproven conjecture that turbulence might be defined as the *breakdown of smooth solutions* of NSE after a finite time. By smooth we mean:

$$p, \mathbf{v} \in C^\infty(\mathbb{R}^3 \times [0, \infty)) . \tag{3.104}$$

Fig. 3.23 One of the first sketches of turbulence *turbolenza* from Leonardo da Vinci [42] which was drawn around 1510. It includes remarkably many details. Reproduced by Permission of Royal Collection Trust/©Her Majesty Elizabeth II 2013

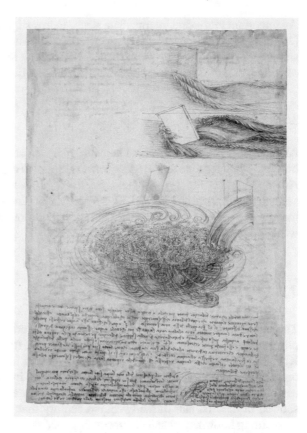

Fig. 3.24 O. Reynolds classical experiments

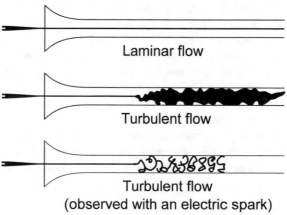

An important tool—and intermediate step—in investigating existence and regularity [19] of NSE is the concept and distinction of *weak* and *strong* solutions. By that the smoothness of the derivatives (particularly important for existence and boundedness of vorticity ω) are tamed to some extent. Strong solutions then are genuine solutions.

3.9.2.2 Dynamical System Approach

Another important tool from mathematical investigations arises from the question: how it is at all possible that a *deterministic system*[4] is capable of becoming *random*.[5] An important concept is that of an *attractor* or *attracting set* in phase space [24]. It seems that within this framework Kolmogorov's picture (see Sect. 3.9.4) of turbulence may be connected to NSE rigorously and, in addition, a specific route or *scenario* was introduced [57].

3.9.3 The Physics of Turbulence

Through the development of the atomic model during the nineteenth century in physics, an improvement in the mechanics of heat (thermodynamics) was gained. Knowing that many ($\sim 10^{23}$) atoms must be present in ordinary amounts of matter, only statistical (averaged) quantities seemed to be of interest. Here statistics emerged as a tool for reduction of information. Even more fundamentally, quantum mechanics showed in the first half of the twentieth century that probabilities not only emerge by omitting known mechanical information, but also as the only way of describing a quantum mechanical system, a hydrogen atom for example. During the Second World War, these methods were generalized to systems of infinite numbers of degrees of freedom (fields), including classical (non-quantum) systems.

It appeared more than natural to apply these methods to the turbulence problem[6] What has been achieved by purely physical methods? A summary of results up to the end of the 1980s may be found in [45]. At that time application of so-called *Renormalization Group (RNG)* methods for derivation of improved engineering turbulence models were heavily debated. A broad variety of theoretical field methods like those presented in the work of R. Kraichnan, e.g. [7] has been tried but with limited success. A notable exception is the exactly analytically solvable model of an advective *passive scalar* [38], sometimes called the *Ising-model* of turbulence [9].

---

[4] Sometimes called chaos theory.

[5] By random we mean that there exist only probabilities for the field quantities, at least in the sense of classical statistical physics.

[6] Many well-known physicists worked on turbulence: W. Heisenberg, L. Onsager, Carl-Friedrich von Weizsäcker, to name a few of them. An often repeated quote is that of Richard Feynman, that *turbulence is the most important unresolved problem in classical physics*.

[7] Especially on 2D turbulence for which he was awarded the Dirac Medal in 2003.

3.9.4 Kolmogorov's Theory

3.9.4.1 Length Scales

Turbulence may be defined in terms of time averages. With

$$< u >:= \int_{t_0}^{t_e} u(t) \, dt \stackrel{!}{=} \int_o^\infty P(v) \cdot v \, dv \qquad (3.105)$$

and $t_0 = 0$ and $t_e = 600\,\text{s}$ in wind-energy we may separate the fluctuation u'

$$u(t) =< u > +u' . \qquad (3.106)$$

Clearly $\sigma^2 :=< u'^2 >\neq 0$. The last equality in Eq. (3.105) is connected to the so-called *ergodicity* hypothesis, which means that time averages and ensemble-averages are equal. In the case of wind this is simply a question of how long to measure to obtain a reliable probability-density function of wind speed. In most practical cases this is reduced to an estimation of two parameters (A, k) of Eq. (1.2). Let P and P' be two points with location $\mathbf{x}$ and $\mathbf{x}'$ and $\mathbf{r} = \mathbf{x} - \mathbf{x}'$. The correlation tensor then takes a special form:

$$< u'_\alpha(\mathbf{x})u'_\beta(\mathbf{x}') >= Q_{\alpha\beta}(\mathbf{r}) =< u'^2 > (f(r) - g(r))\frac{r_\alpha r_\beta}{r^2} + < u'^2 > g(r) \cdot \delta_{\alpha\beta}$$
$$(3.107)$$

only if the flow is homogenous and isotropic [3]:

- Homogeneity: correlations only depend on $\mathbf{r}$ and not on $\mathbf{x}, \mathbf{x}$'
- Isotropy: no direction is preferred.

With that at least three length scales may be defined:

1. an integral correlation length L
2. a micro-scale of *Taylor* λ_T
3. a dissipation length of *Kolmogorov* η_L

with (Fig. 3.25).

$$L := \int_0^\infty f(r)dr \qquad (3.108)$$

$$\frac{1}{\lambda_T^2} := -f''(0) \qquad (3.109)$$

Equation (3.109) implies that λ_T may be easily derived from a Fourier Transform of a wind time series (see Fig. 3.26).

$$\rho(s) :=< u(0) \cdot u(s) > / < u'(0)^2 > \qquad (3.110)$$

Fig. 3.25 Global and Taylor
length scales

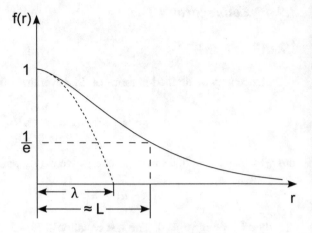

Fig. 3.26 Example of an
autocorrelation, Eq. (3.110).
Data from [37]. Note that the
slope at t = 0 appears to be
non-zero because of
violation of non-stationarity

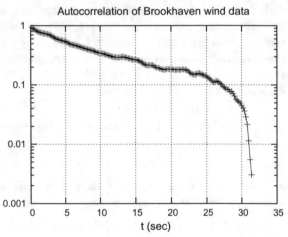

Originally Taylor thought that λ_T gives the size of the smallest eddies within a
turbulent flow. This is not the case, however, as Kolmogorov showed.

With (3.21) for the kinetic energy within a volume V it can easily be shown from
NSE at $\rho \equiv 0$ that

$$\frac{dE_k}{dt} = \int_V u_\alpha \cdot f_\alpha - \int_V \varepsilon \, dV \tag{3.111}$$

introducing

$$2 \cdot \varepsilon = \nu \left(u_{\alpha,\beta} + u_{\beta,\alpha}\right)^2 = \nu \cdot (\nabla \times \mathbf{v})^2 \tag{3.112}$$

and using again a notation of Einstein:

$$F_{\alpha,\beta} := \frac{\partial F_\alpha}{\partial x_\beta}. \tag{3.113}$$

λ_T may be used to connect [54] u'^2 and ε:

$$\varepsilon = 15\nu \frac{<u'^2>}{\lambda_t^2} . \tag{3.114}$$

In flows with no obvious length-scale, a turbulent Taylor–Reynolds number is often used:

$$Re_{\lambda_T} := \sqrt{<u'^2>} \cdot \lambda_T/\nu \tag{3.115}$$

3.9.5 Dissipation Scales

+ Kolmogorov introduced in 1941 [22, 54] a theory of turbulence with a very small set of assumptions and—remarkably—without reference to the NSE. He assumed that at *sufficiently high Reynolds-numbers* only two parameters ε (dissipation) and ν (kinematical viscosity) determine all the properties of the flow. As we already know, $\nu \sim 10^{-5}$ for air and ε may be estimated via $\sim u'^3/L$ From that scales for length, time and velocity may be defined as follows:

$$\eta = (\nu^3/\varepsilon)^{1/4}, \tag{3.116}$$
$$\tau_\eta = (\nu/\varepsilon)^{1/2}, \tag{3.117}$$
$$u_\eta = (\nu \cdot \varepsilon)^{1/4} . \tag{3.118}$$

At length scales less then η, dissipation into heat starts. Therefore the ratio

$$L/\eta \sim Re^{3/4} \tag{3.119}$$

gives an estimate of the number of points to be resolved in a numerical solution of the NSE. In 3D and assuming a $Re_L \sim 10^6$ we then arrive at $N_{points} \sim 3 \cdot 10^{13}$. Resolving turbulent flow down to this scale is called *Direct Numerical Simulation (DNS)* . In [28] such a HIT (= **H**omogeneous **I**sotropic **T**urbulence) Simulation is reported for λ_T-values up to 1200.

From Table 3.3 a rough estimate for η from wind measurement is shown. It gives values on the order of 1 mm.

Another important parameter is the kinetic energy density as well as the standard deviation. Both have the dimension $(m/s)^2$.

Defining the *turbulent kinetic energy* by

$$k := \frac{1}{2} <u_i' \cdot u_i'> \tag{3.120}$$

we may look for the energy contained in a specific wave-number range,[8] Slightly generalizing Eq. (3.110)

$$R_{ij}(\mathbf{r}, \mathbf{x}, t) := < u_i'(\mathbf{x}, t) \cdot u_i'(\mathbf{x} + \mathbf{r}), t > \tag{3.124}$$

we now transform to

$$\Phi_{ij}(\kappa, t) := \frac{a}{(2\pi)^3} \int e^{-i\kappa \cdot \mathbf{r}} R_{ij}(\mathbf{r}, \mathbf{x}, t) \cdot d\mathbf{r} . \tag{3.125}$$

The *energy spectrum* function now reads

$$E(\kappa, t) := \frac{1}{2} \int \Phi_{ii}(\kappa, t) \delta(|\kappa| - \kappa) d\kappa . \tag{3.126}$$

It may easily shown that

$$\int_o^\infty E(\kappa, t) d\kappa = \frac{1}{2} R_{ii}(0, t) = \frac{1}{2} < u_i \cdot u_i > . \tag{3.127}$$

Now with the same class of arguments Kolmogorov showed

$$E(\kappa) = \varepsilon^{2/3} \cdot \kappa^{-5/3} \cdot \Psi(\kappa \cdot \eta) . \tag{3.128}$$

Figure 3.27 shows measurement [30] at a site close to the North Sea indicating validity of Kolmogorov's *five–thirds law* for at least five orders of magnitude.[9]

In wind energy, semi-empirical spectra have to be used for generating *synthetic wind*. Equation 3.130 and Fig. 3.28 show one commonly used, originating from von Karman.

$$E(\kappa) = C_K \cdot \varepsilon^{2/3} \cdot \kappa^{-5/3} \cdot f_L(\kappa \cdot L) f_\eta(\kappa \cdot \eta) \tag{3.129}$$

$$f_L(\kappa \cdot L) = \left(\frac{\kappa L}{[(\kappa L)^2 + c_L]^{1/2}} \right)^{5/3 + n_0} . \tag{3.130}$$

---

[8] It is well known from the theory of linear partial differential equations that a number of problems are simplified if formulated by Fourier transform into wavenumber–frequency co-space. The NSE equation then reads as

$$k_\beta \cdot u_\beta(\mathbf{k}, t) \tag{3.121}$$

$$\left(\frac{\partial}{\partial t} + v \cdot k^2 \right) u_\alpha(\mathbf{k}, t) = M_{\alpha\beta\gamma}(\mathbf{k}, t) \sum u_\beta(\mathbf{j}, t) \cdot u_\gamma(\mathbf{k} - \mathbf{j}, t) \text{ and} \tag{3.122}$$

$$M_{\alpha\beta\gamma}(\mathbf{k}, t) = (2i)^{\mp 1} \left(k_\beta D_{\alpha\gamma}(\mathbf{k}) + k_\gamma D_{\alpha\beta}(\mathbf{k}) \right) . \tag{3.123}$$

[9] Taylor's frozen turbulence hypothesis has been used here. It states that time series may be used instead of spatially varying values.

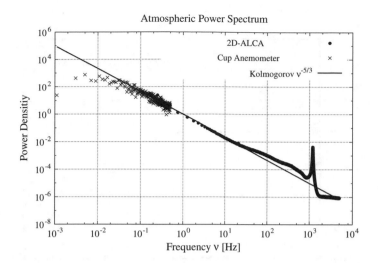

Fig. 3.27 Energy density spectra close to the North Sea. The peak at 1 kHZ is an artifact of the measurement system. 10^{-2} Hz seems to be an approximate maximum

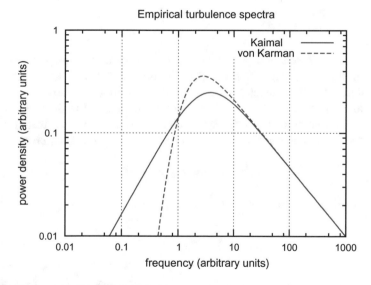

Fig. 3.28 Semi-empirical turbulence spectra

3.9.6 Turbulence as a Stochastic Process

Stochastic Processes [33] were first used in description of *Brownian motion* , where the motion of particles is described by classical mechanics and is subjected to additional *random forces*. Of course, specifics must be applied to define the meanings and differences of *random*, *statistical*, and *stochastic*. As *Statistical Fluid Mechanics* [49]

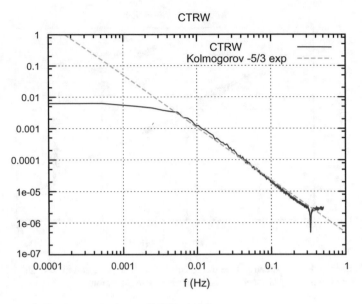

Fig. 3.29 Energy density spectra from a CTRW-model

often refers to the *Mechanics of Turbulence*, it seems natural to apply these methods to turbulent flows. Starting with the work of Friedrich and Peinke [23] attempts were made to derive Fokker–Planck Equations for the description of turbulence. We will discuss this important issue of *synthetic wind* in more detail in Sects. 4.5 and 4.6, see [35, 60]. As an outcome of this approach (see Fig. 3.29) we see that at least the important starting points are *velocity increments* and *structure functions*:

$$\delta \mathbf{u}(\mathbf{x}, \mathbf{r}) := \mathbf{u}(\mathbf{x} + \mathbf{r}) - \mathbf{u}(\mathbf{r}) \tag{3.131}$$

which according to Kolmogorov's theory [22] scales like

$$< (\delta \mathbf{u} r)^2 > = C \cdot \varepsilon^{2/3} \cdot r^{2/3} \tag{3.132}$$

$$< (\delta \mathbf{u} r)^3 > = -\frac{4}{5} \cdot \varepsilon \cdot r \tag{3.133}$$

$$S_n(r) = < (\delta \mathbf{u} r)^n > = C_n \cdot \varepsilon^{n/3} \cdot r^{n/3} . \tag{3.134}$$

An interesting parameter called *intermittency* originally means the occurrence of intermittent bursts of high activity. In addition it may cause a deviation of the Kolmogorov scaling $log(S_n) \sim n$ [22] and may serve as a model of *gusts* [6] (Fig. 3.30).

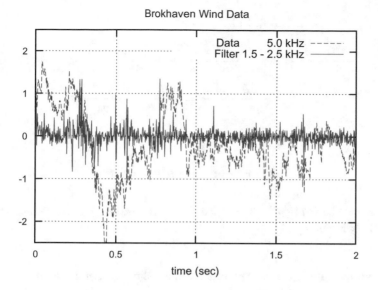

Fig. 3.30 Intermittent time signal from high-pass filtering of wind data

3.9.7 The Turbulent Boundary Layer

Boundary layer theory successfully describes the velocity profile coming from the outer inviscid flow to the wall. In its simplest form it is the Blasius profile for a flat plate. If turbulence comes into play, unfortunately things become much more complicated even for a simple flat plate [54].

Nevertheless this type of flow is important both in application to wind (as will be seen in the next section) and to CFD (= numerically solved Navier–Stokes Equations).

With the term *fully developed* we loosely define a type of a flow in which at all locations large fluctuations (over time) are present. We define *Reynolds averaging (RA)* as separating averages from fluctuations

$$u(t) = U + u'(t) \tag{3.135}$$

with

$$U :=< u(t) >:= \int_{-\infty}^{\infty} u(t) \, dt \,, \tag{3.136}$$

Therefore $< u' >= 0$. Now we consider a simple 2D planar flow. Applying RA and BL-approximations to the appropriate NAST we get

$$\frac{\partial <U>}{\partial x} + \frac{\partial <V>}{\partial x} = 0,$$ (3.137)

$$<U>\frac{\partial <U>}{\partial x} + <V>\frac{\partial <U>}{\partial y} = \nu\frac{\partial^2 <U>}{\partial y^2} - \frac{\partial}{\partial y}<u'\cdot v'>$$ (3.138)

$$\frac{1}{\rho}\frac{\partial <p>}{\partial x} = \frac{1}{\rho}\frac{dp_0}{dx} - \frac{\partial <v'^2>}{\partial x}$$ (3.139)

The shear stress then becomes

$$\tau = \rho\nu\frac{\partial <U>}{\partial y} - \rho<u'\cdot v'> .$$ (3.140)

Expressions like $<u'_i \cdot u'_j>$ are called *Reynolds stresses*. The *wall shear stress* is particularly important

$$\tau_w := \tau(0) .$$ (3.141)

Using this quantity as the primary one we may define a *friction velocity*

$$u_\tau := \sqrt{\frac{\tau_w}{\rho}} .$$ (3.142)

and a *viscous length scale* [54]

$$\delta_\nu := \nu\sqrt{\frac{\rho}{\tau_w}} .$$ (3.143)

Wall distance and velocity now may be measured with reference to these values using dimensionless numbers only

$$y^+ := \frac{u_\tau y}{\nu},$$ (3.144)

$$u^+ := \frac{U}{u_\tau}.$$ (3.145)

3.9.8 The Log Law of the Wall

Boundary layer theory successfully describes the velocity profile for laminar flow coming from outer inviscid flow to the wall. In its simplest form, this is the Blasius profile for a flat plate, as a no-slip boundary condition. If turbulence comes into play, unfortunately things become much more complicated even for a simple flat plate [54]. A law of the wall now is an equation of the type

$$u^+ = f_w(y^+) .$$ (3.146)

In total we may distinguish about 5 [54] sublayers within the turbulent boundary layer, but two are of particular importance:

(a) Very close to the wall ($y^+ < 5$) we must have

$$f_w(y^+) = y^+ . \tag{3.147}$$

Because $u^+(0) = 0$ (no-slip condition), and $f'_w(0) = 1$ because of the special normalization we used. Sometimes this sublayer is called *viscous sub layer*.

(b) Now if $y^+ > 100$ Prandtl and von Karman [54] assumed that viscosity is of no importance and

$$\frac{du^+}{dy} = const. \cdot \frac{u_\tau}{y} \tag{3.148}$$

simply because of dimensional analysis. The constant $const. := \kappa \approx 0.4$ is called *von Karman constant*. If we integrate Eq. (3.148) we get the famous *logarithmic law of the wall*:

$$u^+ = u_\tau \left(\frac{1}{\kappa} \ln(y) + C \right) . \tag{3.149}$$

Figure 3.31 shows an example from the German offshore measurement station FINO1, about 40 km north of the island of Borkum. All data up to 90 m seem to fit rather well to the logarithmic law, with the exception of the highest anemometer at H = 100 m above sea level.

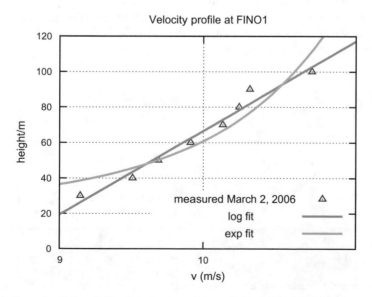

Fig. 3.31 Example of the validity of the log law of the wall. Data from FINO1 [15]

3.10   Wind as Turbulent Flow in the Lower Atmosphere

The methods applied above may also be applied to the *Atmospheric Boundary Layer* (ABL) , which is sometimes also called *Planetary Boundary Layer*. Two important distinctions have to be noted:

- geometry is that of the surface of a sphere;
- there is no inertial system of reference, the earth is rotating with $\omega_{earth} = 7 \cdot 10^{-5}/\text{s}$.

The outer flow is called *geostrophic wind* and is induced by the combined action of pressure-gradients and Coriolis forces.

A naive estimation assuming laminar flow produces far too small values for the thickness of the ABL of only few meters [66]. More realistic estimations must include effects from turbulence and thermal radiation. Putting everything together, one can say that the thickness of the ABL varies from approximately very few hundreds of meters (at night) to about 1500 m in the daytime (Fig. 3.32).

One special mention has to be made of *roughness length* Eq. (3.149), which is used for estimating wind velocities at different heights, when one measurement z_r, v_r is known (Table 3.2):

$$v(z) = v_r \cdot \frac{ln(z) - ln(z_0)}{ln(z_r) - ln(z_0)} \ .$$ (3.150)

Obviously $v(z_0) \equiv 0$.

We will learn more in Sect. 4.2 about vertical profiles.

As already shown, wind may be used as an example to show the validity of the Kolmogorov theory across more than five decades (Fig. 3.27). Figure 3.33 shows an example of the underlying time series.

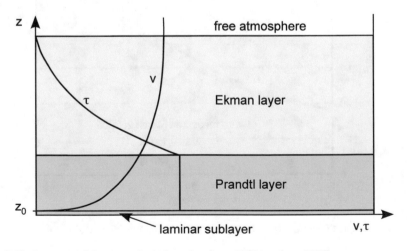

Fig. 3.32 Structure of the atmospheric boundary layer (ABL), adapted [66]

Table 3.2 Some examples of roughness heights

Surface	z_0 (m)
Water	0.1–1 10^{-3}
Shrub	0.1–0.2
Forest	0.5
Cities	1–2
Mega cities	5
Mountains	1–5

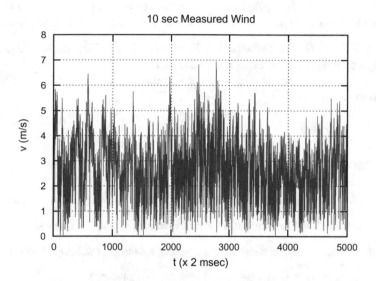

Fig. 3.33 Typical time series of a high-frequency wind speed measurement

Table 3.3 Measurements of turbulent dissipation rates in atmospheric winds

Author	Source	ε	$\sqrt{<u'^2>}$ (m/s)
Kunkel et al.	[36]	0.1 to $1 \cdot 10^{-2}$	~1.0
Srinivasan et al.	[37]	$1.0 \cdot 10^{-2}$	~1.5
Jeromin, Schaffarczyk	[30]	$\sim 3 \cdot 10^{-2}$	~2.0

Many more turbulence characteristics may be derived from these types of measurements. Table 3.3 shows comparison of the turbulent dissipation rate in wind (from [30, 36, 37]).

3.11   Problems

Problem 3.1 Derive Bernoulli's equation from the Euler Equation, Eq. (3.31).

Problem 3.2 Derive the velocity distribution of a line vortex from integration and calculate the kinetic energy contained.

Problem 3.3 Use [3] to find the complex velocity potential flow of (Fig. 3.7, right) and calculate the dividing (free) streamline.

Problem 3.4 Use [52] or [48] to investigate the Joukovski transformation (3.51) in more detail. Find an expression for the thickness and the camber as function of z_m.

Problem 3.5 (a) Derive from Table 3.3 corresponding values for the Taylor- and Kolmogorov length scales, Eqs. (3.114) and (3.116).

(b) Discuss your results with reference to the size of the active measurement equipment.

References

1. Abbott I, von Doenhoff A (1959) Theory of wing sections: Including a summary of airfoil data. Dover, Mineola
2. Bachelor GK (1967) An introduction to fluid dynamics. Cambridge University Press, Cambridge
3. Bachelor GK (1993) The theory of homogeneous turbulence. Cambridge University Press, Cambridge
4. Babinsky H (2003) How do wings work? Phys Educ 8(6):497–503
5. Boyd JP (1999) The Blasius function in the complex plane. Exp. Math. 8:1–14
6. Böettcher F, Barth St, Peinke J (2006) Small and large scale fluctuations in atmospheric wind speeds. Stoch Environ Res Risk Asses
7. Blasius H (1908) Grenzschichten in Füßigkeiten mit kleiner Reibung. Z Math Phys 66:1–37 in German
8. Cebeci T, Cousteix J (1999) Modeling and computation of boundary-layer flows. Springer, Berlin
9. Chen S et al (2008) Obituary of R. Kraichnan. Phys Today 71:70–71
10. Chorin A (1993) A mathematical introduction to fluid mechanics. Springer, New York
11. Chorin A (1994) Vorticity and turbulence. Springer, New York
12. Constantin P (2007) On the Euler Equations of incompressible Fluids. Bull Am Math Sci 44:603–621
13. Darrigol O (2005) Worlds of flow. Oxford University Press, Oxford
14. Drela M (1989) XFOIL: an analysis and design system for low reynolds number airfoils. Springer lecture notes in engineering, vol 54. Springer, Berlin, pp 1–12 (1990)
15. Emeis S (2012) Private communication
16. Emeis S (2013) Wind energy meteorology. Springer, Berlin
17. Eppler R (1990) Airfoil design and data. Springer, Berlin
18. Eckert M (2006) The dawn of fluid mechanics. Wiley-VCH, Weinheim
19. Fefferman C (2000) Existence & Smoothness of the Navier-Stokes Equations. Clay Mathematics Institute, The Millennium Prize, Problems, Navier-Stokes Equations, Providence, RI, USA

20. Foias C, Manley O, Rosa R, Temam R (2001) Navier-Stokes equations and turbulence. Cambridge University Press, Cambridge
21. Föppl O (1911) Windkräfte an ebenen und gewölbten Platten, Jahrbuch der Motorluftschiff-Studiengesellschaft, Berlin, Germany (in German)
22. Frisch U (1995) Turbulence. Cambridge University Press, Cambridge
23. Friedrich R, Peinke J (1997) Description of a turbulent cascade by a Fokker-Planck equation. Phys Rev Lett 78:nn mm
24. Gallavotti G (2002) Foundations of fluid mechanics. Springer, Berlin
25. Glauert H (1926) The elements of aerofoil and airscrew theory, Repr, 2nd edn. Cambridge University Press, Cambridge
26. Hansen MOL (2008) Aerodynamics of wind turbines, 2nd edn. Earthscan, London
27. Hermandez GGM (2011) Laminar-turbulent transition on wind turbines. Technical University of Copenhagen, Denmark Phd Thesis
28. Ishihara T, Gotoh T, Kaneda Y (2009) Study of high-reynolds number isotropic turbulence by direct numerical simulation. Annu Rev Fluid Mech 41:165–180
29. Jacobs M (2015) High reynolds number airfoil test in DNW-HDG, DNW-GUK-2014 C04. Göttingen, Germany July
30. Jeromin A, Schaffarczyk AP (2012) Advanced statistical analysis of high-frequency turbulent pressure fluctuations for on- and off-shore wind. In: Proceedings of Euromech coll 528, Oldenburg, Germany
31. Jones RT (1990) Wing theory. Princeton University Press, Princeton
32. Kant I (1786) Metaphysische Anfrangsgründe der Naturwissenschaft. Königsberg, Königreich Preussen
33. van Kampen NG (2007) Stochastic processes in physics and chemistry, 3rd edn. Elsevier, Amsterdam
34. Katz J, Plotkin A (2001) Low-speed aerodynamics, 2nd edn. Cambridge University Press, Cambridge
35. Klein F (1910) Über die Bildung von Wirbeln in reibungslosen Flüssigkeiten. Z f Mathematik u Physik 58:259–262
36. Kunkel KE, Eloranta EW, Weinman JA (1980) Remote determination of winds, turbulence spectra and energy dissipation rates in the boundary layer from lidar measurements. J Atmos Sci 37(6):978–985
37. Kurien S, Sreenivasan KR (2001) Measures of anisotropy and the universal properties of turbulence. In: Lesieur M et al (eds) Les Houches summer school in theoretical physics, session LXXIV. Springer, Berlin
38. Kraichnan R (1994) Anomalous scaling of a randomly advected passive scalar. Phys Rev Lett 72:1016–1019
39. van Kuik G (2018) The fluid dynamic basis for actuator disc and rotor theories. IOS Press BV (open access). https://doi.org/10.3233/978-1-61499-866-2-i
40. Lamb SH (1932) Hydrodynamics, 6th edn. Cambridge University Press, Cambridge
41. Leishman JG (2002) Challenges in modeling the unsteady aerodynamics of wind turbines, AIA 2002–0037. Reno, NV, USA
42. Leonordo da Vinci, Verso: studies of flowing water c. 1510–1512, with notes, RL 12660v. Royal Collection Trust, Windsor, UK
43. Lortz D (1993) Hydrodynamic. B.I.-Wissenschaftsverlag, Mannheim
44. Lions P-L (1996) Mathematical topics in fluid mechanics: volume 1: incompressible models. Clarendon Press, Oxford
45. McComb WD (1992) The physics of fluid turbulence. Clarendon Press, Oxford
46. Meier GEA, Sreenivasan KR (eds) (2006) IUTAM symposium on one hundred years of boundary layer research. Springer, Berlin
47. Majda AJ, Bertozzi AL (2002) Vorticity and incompressible flow. Cambridge University Press, Cambridge
48. Milne-Thomson LM (1996) Theoretical hydrodynamics, 5th edn. Dover Publications, New York

49. Monin AS, Yaglom AM (2007) Statistical fluid mechanics, vol 2. Dover Publications, New York
50. Munk MM (1992) General theory of thin wing sections, NACA, Report No. 142
51. Oboukhov AM (1962) Some specific features of atmospheric turbulence. J Fluid Mech 13:82–85
52. Panton RL (1996) Incompressible flow, 2nd edn. Wiley, New York
53. Pereira R, Schepers G, Pavel MD (2013) Validation of the Beddoes-Leishmann dynamic stall model for horizontal axis wind turbines using MEXICO data. Wind Energy 16(2):207–219
54. Pope SB (2000) Turbulent flows. Cambridge University Press, Cambridge
55. Prandtl L, Betz A (2010) Vier Abhandlungen zur Hydrodynamik und Aerodynamik. Universitätsverlag Göttingen, Germany (in german)
56. van Rooij RPJOM (1996) Modification of the boundary layer calculation in RFOIL for improved airfoil stall prediction, Internal report IW-96087R, TU Delft, The Netherlands
57. Ruelle D, Takens F (1971) On the nature of turbulence. Commun Math Phys 20:167–192
58. Ruelle D (1983) Five turbulent problems. Physika 7D:40–44
59. Saffman PG (1992) Vortex dynamics. Cambridge University Press, Cambridge
60. Schaffarczyk AP et al (2010) A new non-gaussian turbulent wind field generator to estimate design-loads of wind-turbines. In: Peinke J, Oberlack M, Talamelli A (eds) Progress in turbulence III. Springer proceedings in physics, vol 131. Springer, Dordrecht
61. Schlichting H, Gersten K (2000) Boundary layer theory. Springer, Berlin
62. Schubauer GB, Skramstad HK (1943) Laminar-boundary-layer oscillations and transition on a flat plate, NACA-TR-909, 1943/47
63. Spalart P (1988) Direct simulation of a turbulent boundary layer up to $r_\theta = 1410$. J Fluid Mech 187:61–98
64. White FM (2005) Viscous fluid flow, 3rd edn. Mc Graw Hill, New York
65. Yates YE (1991) A unified viscous theory of lift and drag of 2-D thin airfoils and 3-D thin wings, NASA report, CR-4414
66. Zdunkowski W, Bott A (2003) Dynamics of the atmosphere. Cambridge University Press, Cambridge

Chapter 4
Inflow Conditions to Wind Turbines

4.1 Importance of Inflow Conditions to Rotor Performance

In Chap. 3 some of the necessary fluid mechanics background was introduced. Now looking closer at the wind turbine, we may ask which details of the turbulent wind field are generally more important and in which order the properties of this field should be discussed. Certainly one may start with the easiest flow field: a spatially and temporally constant one. This model is used in most analytical studies.

Nevertheless, wind is as it is, so we will approach reality in steps of increasing sophistication. Starting with the simplest spatial inhomogeneity, we consider at first a gradual vertical increase of wind speed called wind shear.

4.2 Wind Shear

As we have already seen in Sect. 3.9.8, and as exemplified in Fig. 3.31, a turbulent boundary layer shows a logarithmic increase of wind speed with height. The slope of this curve, described in Eq. (3.150), uses one parameter only, the roughness length. At least two measurements and two different heights are, therefore, necessary for extrapolation to other heights. Figure 4.3 shows an example for a location in Hamburg, Germany. Any decision on hub heights and the resulting tower investment rely heavily on these data (see Table 3.2 and [21]). Figures 4.1 and 4.2 give two more examples for locations in cities where z_0 is particularly large, giving steep velocity gradients.

Due to the change (in most cases decrease) in temperature with height, there is an energy flux in upward direction which influences the profile of the mean horizontal wind speed and the turbulence characteristics. A (Monin–Obukhov) length scale

$$L := \frac{u^{\star 3} c_p \rho T}{k g H} \qquad (4.1)$$

© Springer Nature Switzerland AG 2020
A. P. Schaffarczyk, *Introduction to Wind Turbine Aerodynamics*,
Green Energy and Technology, https://doi.org/10.1007/978-3-030-41028-5_4

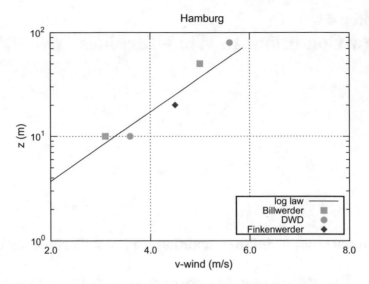

Fig. 4.1 Wind profile at locations (Billwerder, Finkenwerder) close to Hamburg, Germany. DWD means German weather service)

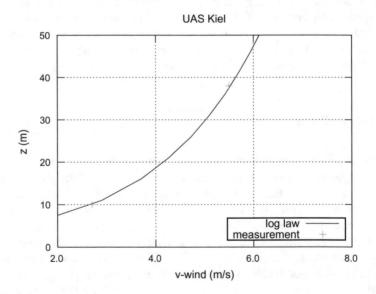

Fig. 4.2 Wind profile at Kiel, Germany

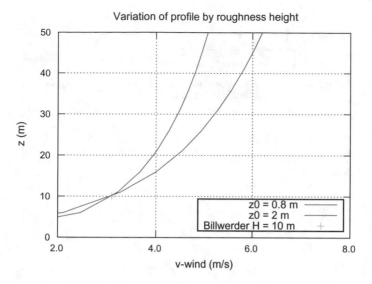

Fig. 4.3 Dependence of wind profile shape (City) on z_0

has to be introduced with $u^\star$ the friction velocity, c_p specific heat of the air, ρ density, T, temperature, k von Karman's constant ≈ 0.4, g earth acceleration, and H the flux. The profile then is expressed [21] as

$$u(z) = \frac{u^\star}{k} \left[ln \frac{z}{z_0} - \psi_m \left(\frac{z}{L} \right) \right] \tag{4.2}$$

with a universal function $\psi_m(z/L)$. As a result the profiles change significantly diurnally, and the profile is fundamentally different over water surfaces, which is important for offshore wind energy applications (see Fig. 4.4).

4.3 Unsteady Inflow Versus Turbulence

As wind turbines are flexible structures, they may be able to react to instationary inflow conditions up to a certain frequency. These limit frequencies are on the order of a few Hertz, as may be seen from a sample *Campbell* Diagram. Therefore from a structural point of view, the transition between instationary flow and turbulence may be drawn (without any rigorous reasoning) at 10 Hz.

From a more aerodynamic standpoint, the situation is not as clear. Starting with a laminar Boundary Layer, Stokes and others [4] have shown, that starting from a simple oscillation flow

$$u(x, t) = u_0(x) \left(1 + B \cdot cos(\omega t) \right) \tag{4.3}$$

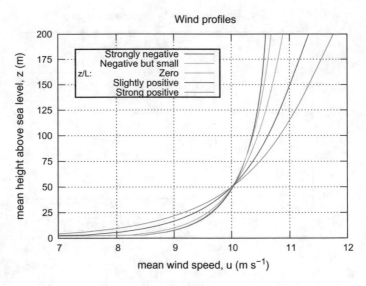

Fig. 4.4 Wind profiles as function of temperature differences

the response is spatially limited due to viscous damping. For the laminar case, the boundary layer is able to respond within a length $\sim 1/k$ perpendicular to the main flow only for

$$k = \sqrt{\frac{\omega}{2\nu}} \, . \tag{4.4}$$

For air and $\omega = 60/s$ we have approx. 4 mm for this length scale.

In turbulent flow there is even a characteristic frequency called the *bursting frequency* [7] which may be defined as

$$f_B = U/5\delta_T \tag{4.5}$$

where δ_T is the turbulent boundary layer thickness. On a wind turbine blade with $U = 40$ m/s we may estimate $\delta_Z \approx 7$ mm and therefore $f_B \approx 200$ Hz. To conclude we may designate a very blurred border between structural and aerodynamic turbulence somewhere between 10 and 200 Hz.

4.4 Measuring Wind

Measuring the wind is an important part during site assessment. As special examples we show in Figs. 4.6, 4.7, 4.8, and 4.9 measurements for storm *Christian* occurring in Northern Germany in October 2013. KWK is a location close to the North sea, whereas FINO3 is located about 80 km west of the German island of Sylt.

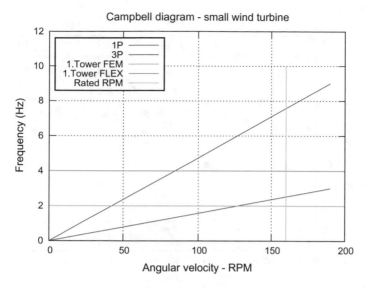

Fig. 4.5 Sample resonance (Campbell) diagram of a small wind turbine. FEM means results from a finite element method code and FLEX is an often used aeroelastic tool

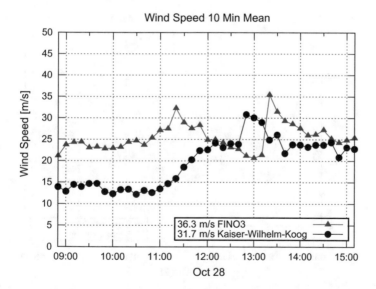

Fig. 4.6 Storm Christian on October 28, 2013, at FINO3 and KWK, 10 min average

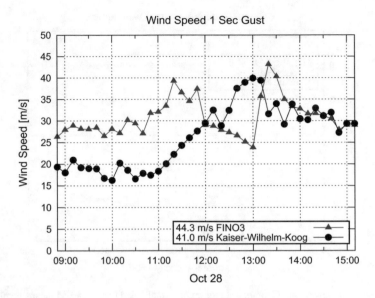

Fig. 4.7 Storm Christian on October 28, 2013, at FINO3 and KWK, 1 s gust

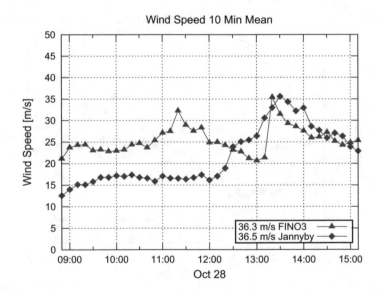

Fig. 4.8 Storm Christian on October 28, 2013, at FINO3 and Janneby, 10 min average

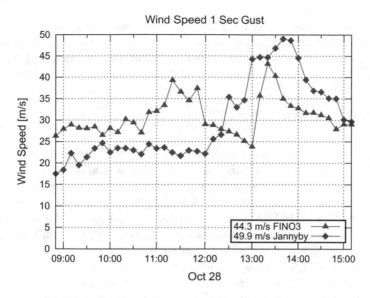

Fig. 4.9 Storm Christian on October 28, 2013, at FINO3 and Janneby, 1 s gust

4.5 Synthetic Winds for Load Calculation

It was recognized very early in the development of modern wind turbines that instationary loads from fluctuating wind may cause very early failure of components due
to *fatigue*.[1] Therefore so-called aeroelastic simulations were initiated, which relied
on artificial or synthetic winds [26]. One of the earliest attempts was made by Veers
[25] and is still often used. A more recent overview may be found in Kleinhans [14].
In principle the path from the model to a synthetic time series is as follows (Fig. 4.5):

- select a model energy spectrum,
- select a model of anisotropic coherence (spatial decay),
- perform Fourier transforms to get time series.

The Veers Model is isotropic, which is not appropriate for wind, as the lateral fluctuation components are much smaller than the others. Mann [17] tries to include this
anisotropy in a much more sophisticated model. In an attempt by Schaffarczyk and
others [22], a model from the theory of stochastic processes (so-called *Continuous
Time Random Walks* was applied to simulate turbulent wind [13]. A comparison was
performed for two state-of-the art multi-MW turbines (3 and 5 MW) and in summary
it may be said that at least 10% deviations in the results for the fatigue loads may be

---

[1]A well known example is the German early Multi-MW turbine GROWIAN (1976–1987) which
had to be shut down after only about 400 h of operation [23].

seen. It was even observed that some completely wrong results appear [24]; therefore, it is clear that many more efforts must be made to get reliable and more sophisticated synthetic wind models that are appropriate for the certification process [18].

4.6 Synthetic Winds for CFD

A similar problem appears if one tries to simulate wind farms on a scale much larger than the typical atmospheric length scales of several hundred meters. Then specific inflow conditions are definitely needed if a special kind of transient calculations (Large Eddy Simulation—LES, see Chap. 7) are performed. An impressive description of the state of the art may be found in [19] (Figs. 4.6, 4.7, 4.8 and 4.9).

In case of turbulent inflow conditions it is necessary that spatially and temporally resolved turbulent flow fields are set up such that they statistically resemble the real-world or experimental conditions as close as possible. This results in the need for setting up the so-called turbulent boundary conditions. There have been many approaches to model these spatiotemporal inflow conditions, some of the early methods can be found in Kleinhans [14].

4.6.1 General Approach: From Experimental Data to Synthetic Time Series

A common approach for the generation of artificial inflow turbulence data that satisfies certain statistical properties known from experimental data such as the energy spectra, mean velocity, fluctuations, cross-correlations, length and time scales, etc., is as follows:

1. For each velocity component, a three-dimensional velocity signal is generated such that the two-point statistics are fulfilled.
2. The cross-correlations between the different velocity components are considered for the generation of the velocity signal in case of anisotropic inflow conditions.
3. An inverse Fourier transform is then used on the obtained velocity signal generating velocity data that satisfies the specified energy spectrum.

To successfully generate time-series data using this approach, one would need three-dimensional energy spectra which is experimentally difficult to obtain. To overcome this problem, Lee et al. [15] uses a model spectrum which represents isotropic turbulence:

$$E(k) \sim k^4 exp(-2(k/k_0)^2) \text{ with wave number vector k}$$
$$k = (k_1^2 + k_2^2 + k_3^2)^{1/2} k_0 = \text{peak wave number.} \tag{4.6}$$

Equation (4.6) is one of the many possible spectra that can be employed. As seen in Klein et al. [12] and is commonly known as the question on how to choose k_0 cannot be uniquely answered. To overcome some of the challenges faced with the general approach outlined above, a method based on digital filtering of random data was developed.

4.6.2 Digital Filters

The bases for the development of the digital filter method by Klein et al. was its practicability. This means that only those statistical quantities that can be obtained with reasonable expenses or from heuristical estimates would be the necessary input. It was concluded that correlation functions and length scales are adequate alternatives.

The flow signal can be described as the sum of the average velocity and a fluctuating part. Using the Cholesky decomposition a_{ij} of the Reynolds stress tensor and synthetically generated velocity fluctuations u_m, Klein et al. use the method by Lund et al. [16] to ensure the correct cross-correlations. The instantaneous velocity signal is as follows:

$$u = \bar{u} + a_{ij}(u_m). \tag{4.7}$$

The definition of the tensor a_{ij} can be found in Lund et al. [16] and Klein et al. [12] To create the two point correlations a series of random numbers r_m is defined such that $\overline{r_m} = 0$ and $\overline{r_m r_m} = 1$, then the convolution or a digital non-recursive filter with filter coefficient b_n and filter length N is defined as seen in Klein et al. [12]

$$u_m = \sum_{n=-N}^{N} b_n r_{m+n}. \tag{4.8}$$

Since $\overline{r_m r_n} = 0$ for $m \neq n$, it follows the autocorrelation:

$$\frac{\overline{u_m u_{m+k}}}{\overline{u_m u_m}} = \sum_{j=-N+k}^{N} b_j b_{j-k} \bigg/ \sum_{j=-N}^{N} b_j^2 \tag{4.9}$$

This gives a relation between the filter coefficients and the autocorrelation function. A 3D filter can be constructed through the product of three 1D filters:

$$b_{ijk} = b_i . b_j . b_k \tag{4.10}$$

Klein et al. proposed in contrast to the full autocorrelation function $R_{uu}(x, r)$ where r is a distance vector and $r = |r|$, a length scale should be prescribed. This leads to a special shape of R_{uu}. With the assumption of a homogeneous turbulence and a fixed time as seen in Batchelor [1]

$$R_{uu}(r, 0, 0) = exp\left(-\frac{\pi r^2}{4L^2}\right). \tag{4.11}$$

With Δx as the grid spacing and $L = n\Delta x$ the length scale, the autocorrelation function in discretized form is

$$\frac{\overline{u_m u_{m+k}}}{\overline{u_m u_m}} = R_{uu}(k\Delta x) = exp\left(-\frac{\pi(k\Delta x)^2}{4(n\Delta x)^2}\right) = exp\left(-\frac{\pi k^2}{4n^2}\right). \tag{4.12}$$

The one-dimensional filter coefficient being

$$b_k \approx \tilde{b}_k \bigg/ \left(\sum_{j=-N}^{N} \tilde{b}_j^2\right)^{1/2} \quad \text{and } \tilde{b}_k := exp\left(-\frac{\pi k^2}{2n^2}\right). \tag{4.13}$$

This by convolution gives the three-dimensional filter as seen in Eq. (4.10). A more efficient procedure to reduce computational effort and memory requirement as well as increase parallel scaling performance is proposed by Kempf et al. [11]. Here an exponential filter kernel is applied to the field of random noise which is generated such that any parallel process may generate the same random number for any given location within the domain and the filter is chosen such that the required integral length scale is recoverable.

4.6.3 Adding Turbulence Close to the Region of Interest

A common problem associated with the introduction of the turbulent boundary conditions at the inflow plane of the computational domain is that the larger grid resolution leads to the damping out of essential higher frequencies. This problem can be overcome by introducing the generated turbulence close to the region of interest which typically has a finer grid resolution. However, care must be taken to ensure that the divergence-free criteria is maintained.

One way to accomplish this is through a source term formulation where an additional source term is added to the momentum equation. A detailed description using the integral form of a general conservation equation as used in the finite volume scheme is found in Schmidt and Breuer [3]. It is shown that despite the superposition of the synthetically generated velocity fluctuations as a source term, the subsequent iterations within their predictor–corrector scheme ensures a divergence-free formulation at the end of the time step.

Figure 4.10 shows the Cartesian velocity components of the instantaneous velocity fluctuations in a plane normal to the inflow and normalized by the inflow velocity for a turbulence intensity of 2.8%. An isotropic case with a length scale per unit chord of

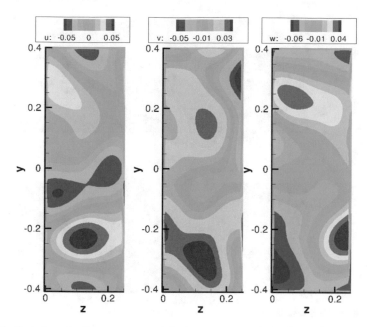

Fig. 4.10 Cartesian velocity components of the instantaneous velocity fluctuations normalized by u_∞ and generated by the digital filter method

0.118 was considered based on an experiment by Hain et al. [8]. Figure 4.11 shows an instantaneous view of the computational domain with and without this added inflow turbulence at a distance of one chord length upstream of the leading edge.

4.7   Problems

Problem 4.1 Imagine a rotor of diameter $\varnothing = 175$ m and hub height $z_1 = 100$ m. Assume a wind speed $u_1 = 12$ m/s and a logarithmic profile with $z_0 = 0.001 \, (0.01)$ m.

Now calculate the following:

(a) (area)averaged wind speed

and

⋆ (b) averaged power

in comparison to the constant values.

Problem 4.2 Turbulence intensity

$$I := \frac{\sigma_v}{\bar{v}} \tag{4.14}$$

may be used for a simple engineering correction [10] of average power when influenced by I.

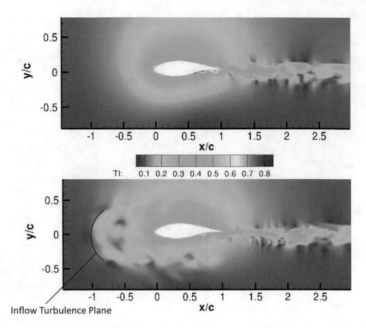

Fig. 4.11 Top: no added inflow turbulence. Bottom: inflow turbulence (TI = 2.8%) introduced one chord length upstream of the leading edge

Show:

$$\overline{P(v)} = P(\bar{v}) + \frac{1}{2} P''(\bar{v}) \sigma_v^2 \ . \tag{4.15}$$

Problem 4.3 Getting familiar with measured wind.

Download 1 h wind data measured with 25 Hz resolution and use some data analysis software (for example, TISEAN from http://www.mpipks-dresden.mpg.de/~tisean/TISEAN_3.0.1/index.html) to get basic information:

(a) Plot a histogram and compare it with a Gaussian distribution.

(b) Calculate the energy spectra which is defined by Eq. (3.126).

c) Finally calculate the (auto)correlation function which is defined by

$$R(\tau) = \frac{1}{\sigma_u^2} \int_0^\infty < u(0) \cdot u(\tau) > \cdot dt. \tag{4.16}$$

Estimate a global correlation time and Taylor's microscale, Eq. (3.109).

Problem 4.4 Gust prediction–return time

(a) From Eq. (5) of [2] a normalized maximum gust wind speed $U_s = \frac{U_{max} - \bar{U}}{\sigma}$ during a large *return time T* may estimated by

$$< U_s >= (2 \cdot ln(\nu T_R))^{1/2} \ . \tag{4.17}$$

Parameter v (not to be mixed up with kinematic viscosity) is related to Taylor's microscale (3.109) by

$$\lambda_T = 1/(v\pi\sqrt{(2)}).\tag{4.18}$$

By using an estimated value for $\lambda_T \approx 20\,\text{ms}$, $\bar{U} = 6\,\text{m/s}$, and $\sigma = 1\,\text{m/s}$ the corresponding gust speed for return times of 1, 10, 50, and 100 years.

(b) Compare to a Rayleigh distribution

$$r(v,\bar{v}) = \frac{\pi}{2}\frac{v}{\bar{v}^2}\cdot\exp-\frac{\pi}{4}\left(\frac{v}{\bar{v}}\right)^2.\tag{4.19}$$

Problem 4.5 Time series averaging.

For typhoon Haiyan from November 2013 wind speed of 65.3 (87.5) m/s during 600 (60) s averaging time were measured. Try to give assumptions for estimation of a 3-s value.

Problem 4.6 (a) Download *TurbSim* from https://wind.nrel.gov/designcodes/preprocesors/turbsim and create sample time series of artificial wind. Produce plot of the time series, a histogram an the energy spectra.

(b) The same with a sample time series from the so-called anisotropic Mann model [11, 12].

* (c) The same with the so-called *continuous time random walk* model [7].

References

1. Batchelor GK (1953) The theory of homogeneous turbulence. Cambridge science classics, Cambridge monographs on mechanics and applied mathematics. Cambridge University Press, Cambridge
2. Beljaars ACM (1987) The influence of sampling and filtering on measured wind gusts. J Atmos Ocean Technol 4:613–626
3. Breuer M, Schmidt S (2017) Source term based synthetic turbulence inflow generator for eddy-resolving predictions of an airfoil flow including a laminar separation bubble. Comput Fluids 146:1–22
4. Cebeci T, Cousteix J (1999) Modeling and computation of boundary-layer flows. Springer, Berlin
5. Cousteix J (1986) Three-dimensional and unsteady boundary layer computations. Annu Rev Fluid Mech 18:173–196
6. Emeis S (2013) Wind energy meteorology. Springer, Berlin
7. Gad-el-Hak M (1989) Feasibility of generating an artificial burst in a turbulent boundary, NAS1-18292, Washington
8. Hain R et al (2009) Dynamics of laminar separation bubbles at low-Reynolds-number aerofoils. J Fluid Mech 630:129–153
9. Kaimal JC, Cifford SF, Lataitis RJ (1989) Effect of finite sampling on atmospheric spectra. Bound-Layer Meteorol 47:337–347
10. Kaiser K et al (2007) Turbulence correction for power curves. In: Peinke J, Schaumann P, Barth S (eds) Wind energy, proceedings of the euromech colloquium. Springer, Berlin

11. Kempf AM et al (2012) An efficient, parallel low-storage implementation of Kleins turbulence generator for LES and DNS. Comput Fluids 60:58–60
12. Klein M et al (2003) A digital filter based generation of inflow data for spatially developing direct numerical or large eddy simulations. J Comput Phys 186(2):652–665
13. Kleinhans D et al (2008) Stochasitische Modellierung komplexer Systeme, Dissertation Universität Münster (in German)
14. Kleinhans D et al (2010) Synthetic turbulence models for wind turbine applications. In: Talamelli A, Peinke J, Oberlack M (eds) Progress in turbulence III. Springer, Dordrecht, pp 111–114
15. Lee S et al (1992) Simulation of spatially evolving compressible turbulence and the application of Taylors hypothesis. Phys Fluids A 4(7):1521–1530
16. Lund TS et al (1998) Generation of turbulent inflow data for spatially-developing boundary layer simulations. J Comput Phys 140(2):233–258
17. Mann J (1998) Wind field simulation. Probab Eng Mech 13(4):269–282
18. NN, IEC 61400, Wind turbines, Design requirements
19. NN (2013) In: Shen WZ (ed) International conference on aerodynamics of offshore wind energy systems, Copenhagen
20. Palutikof JP et al (1999) A review of methods to calculate extreme wind speeds. Meteorol Appl 6:119–132
21. Panofsky HA, Dutton JA (1984) Atmospheric turbulence. Wiley-Interscience, New York
22. Schaffarczyk AP et al (2010) A new non-Gaussian turbulent wind field generator to estimate design-loads of wind-turbines. In: Peinke J, Oberlack M, Talamelli A (eds) Progress in turbulence III. Springer proceedings in physics, vol 131. Springer, Dordrecht
23. Seeger T, Köttgen V, Oliver R (1990) Schadensuntersuchung GROWIAN: Schlussbericht, FB-7-1990,0328949A, Darmstadt
24. Steudel D (2007) Private communication
25. Veers PS (1988) Three-dimensional wind simulation, SAND88-0152, Albuquerque, NM, USA
26. Veltkamp D (2006) Chances in wind energy, a probabilistic approach to wind turbine fatigue design. PhD thesis, TU Delft, The Netherlands

Chapter 5
Momentum Theories

In the meantime, several relevant textbook have been published:

- Reference [38] summarizes advances in momentum theory with special emphasis to connection to vortex theory and compares Joukovskie's and Betz' Vortex model
- Reference [5] starts from *vorticity-based methods* and gives a very detailed compilation of applications to wind turbine aerodynamics
- Reference [19] discusses the path from basic (Euler's) differential equations to the Actuator Disk model
- Reference [8] in Chap. 1 this most recent book gives a *self-contained basis* for design. Special emphasis is given to various approaches to aerodynamical design.

5.1 One-Dimensional Momentum Theory

5.1.1 Forces

We now apply our knowledge from Chap. 3 to the situation shown in Fig. 3.1. To make it quantifiable by hand calculations, we start by using integral momentum theory. We choose three special locations and assume pure one-dimensional flow (in X-direction: u) with no spatial or temporal variation.

- 1: far upstream: In, u_1,
- 2: the location of the wind turbine, Disk, u_2,
- 3: far downstream, Out, u_3.

We assume that the action of the turbine (now also called *actuator disk*) reduces the inflow wind $v_{wind} = u_1$ to u_2 at the disk and further to u_3 far downstream. That there is a further decrease after the disk seems to be somewhat surprising, but this will become clear after proper application of all three theorems of conservation.

Starting with the momentum theorem we see that at *In* we have momentum flow $\dot{p}_1 = \dot{m} \cdot u_1$, whereas at *out* it is only $\dot{p}_3 = \dot{m} \cdot u_3$. From that we have to conclude

© Springer Nature Switzerland AG 2020
A. P. Schaffarczyk, *Introduction to Wind Turbine Aerodynamics*,
Green Energy and Technology, https://doi.org/10.1007/978-3-030-41028-5_5

that there must be a force T exerted by the turbine on the passing air, and from Newton's Third Law, we know there is an opposing force of the same magnitude $-T$ applied by the air to the disk. Together

$$\dot{m}(u_1 - u_3) - T = 0. \tag{5.1}$$

Now we use the Energy Equation (Bernoulli's law). Because we have seen that there is a force at location 2, this may be converted to a pressure drop $\Delta p_{2-2^+} = T/A_r$. As already remarked the reference area A_r is the *swept area* of the rotor $\frac{\pi}{4}D^2$. That implies that we may use Bernoulli only from $1 \rightarrow 2^-$ and then further from $2^+ \rightarrow 3$. Here $-$ refers to a location immediately upstream of the disk and $+$ to a location immediately downstream of the disk. From $1 \rightarrow 2^-$:

$$p_0 + \frac{\rho}{2}u_1^2 = p_2^- + \frac{\rho}{2}u_2^2. \tag{5.2}$$

and the same from $2^+ \rightarrow 3$:

$$p_2^+ + \frac{\rho}{2}u_1^2 = p_0 + \frac{\rho}{2}u_3^2. \tag{5.3}$$

Figure 5.1 shows the development of the pressure along the slipstream. As there is a pressure drop across the disk to negative values (if $p_0 = 0$), there must be a further decrease of velocity for recovering the pressure to p_0.

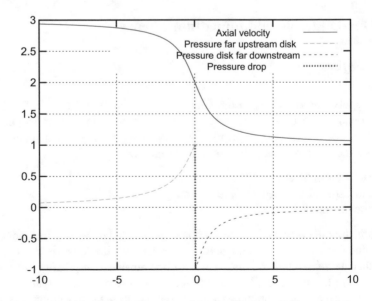

Fig. 5.1 Development of velocity and pressure along the slipstream (arbitrary units)

p_0 denotes the ambient pressure and may be set to 0 without loss of generality. We assume that it acts on the whole surface of our *control volume*.

Solving Eqs. (5.2) and (5.3) for $\Delta p = p_2^+ - p_2^-$ we get

$$\Delta p = \frac{\rho}{2} \left(u_1^2 - u_3^2 \right) . \tag{5.4}$$

Now we have to find an expression for the mass flow which must be constant from $1 \rightarrow 2 \rightarrow 3$. Taking the definition of the density into account and assuming—as already mentioned—constant velocity across A_r we have

$$\dot{m} = \rho \dot{V} = \rho A_r \cdot u_2 . \tag{5.5}$$

Equivalently this equation may be regarded as the *definition* of averaged velocity at the disk. Comparing now Δp from Eq. (5.1) with that from Eq. (5.4) we arrive at the famous *Froude's Law*:

$$u_2 = \frac{1}{2} \left(u_1 + u_3 \right) . \tag{5.6}$$

If we define a reference force as the product of the stagnation pressure of the wind $q_0 := \frac{\rho}{2} u_1^2$ times our reference area A_r, $F_{ref} := q_0 \cdot A_r$ and a so-called *(axial) induction factor* by $u_2 := (1 - a) u_1$ then it is immediately seen that if $c_T := T/F_{ref}$ we may express nondimensional thrust by only one variable a.

$$c_T(a) = 4 (1 - a) a . \tag{5.7}$$

A slipstream can only be existing if $u_3 \geq 0$. Therefore, the inequality $0 \leq a \leq 1$ must hold.

5.1.2 Power

As stated earlier in Eq. (2.1) the extracted power is simply

$$P_T = P_{extracted} = P_{in} - P_{Out} + \Delta p \cdot \dot{V} = \frac{\rho}{2} A_r \left(u_1^3 - u_3^3 \right) . \tag{5.8}$$

In nondimensional form:

$$c_P(a) = 4a (1 - a)^2 . \tag{5.9}$$

Figure 5.2 shows the relationship of Eqs. (5.9) and (5.7). Note that momentum theory is strictly valid only for $0 \leq a \leq 0.5$. In some cases it is useful to extend the range of $c_T(a)$ to values $a \approx 1$. For that a maximum value of $c_T(a = 1) = 2$ is assumed and a starting point on Eq. (5.7) is usually around $a = 0.3$. The maximum of Eq. (5.9) we differentiate with respect to a and get the famous *Betz–Joukovski* [3, 22] limit.

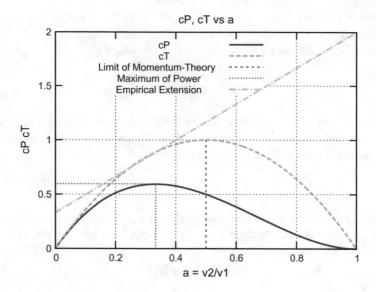

Fig. 5.2 c_P and c_T as function of induction factor a

$$c_P^{max} = \frac{16}{27},$$ (5.10)

$$at\ a = \frac{1}{3},$$ (5.11)

$$and\ c_T\left(a = \frac{1}{3}\right) = \frac{8}{9}.$$ (5.12)

5.1.3   Remarks on Beating Betz

There is an ongoing discussion about whether it might be possible to get closer to a c_P^{max} of 1. In a lesser-known paper—probably because it is published in German only—Betz [4] discussed already in 1926 the possibility of exceeding the $\frac{16}{27}$ value if one relaxes the underlying assumptions, the most restrictive of which is the 2D actuator disk.[1] See Problem 5.2 for a short introduction to the matter.

---

[1]Page 14: Because the ratio $\frac{L_{max}}{L_0}$ plays an exceptionally important role, it is appropriate to consider the extent to which the derived value of 16/27 depends on specific assumptions and if it is possible to increase it intentionally. The energy behind the first wind turbine can easily be exploited by use of a second wind turbine. Our original derivation applies only to a single disk-shaped wind turbine. If one allocates the space behind the first wind turbine for energy use, it is possible to achieve $L_{max/L_=}\approx1$, assuming the prescribed diameter is not exceeded. (translated by John Thayer).

This is particularly obvious in the case of more 3D-like wind-turbines known as DAWTs, see Sect. 2.5. Here most of the confusion comes from the fact that an appropriate definition of the reference area is less obvious. Jamieson [14, 16] proposed the furthest upstream sectional area A_1 to be used as a reference. He further showed that

$$c_P \leq \frac{16}{27}(1 - a_0) \tag{5.13}$$

if a_0 is defined by $v_2 = v_1(1 - a_0)$ when there is no energy extraction from the diffuser—the empty case. It is clear that such a device should have a_0 negative with low additional costs compared to the original turbine.

It is clear then now the discussion can be limited to the diffuser only. (Compare to Ref. [5] from Chap. 2).

5.2 General Momentum Theory

The 1D model of the preceding chapter is a very simple one and may be regarded as purely academic, as it was shown that such a 1D model does not exist at all [37]. One reason for this it that the DEQ of fluid mechanics are coupling the velocity components so rigidly, that from $v \equiv w \equiv\equiv 0$ it must follow $u = const$. In addition no hints on how to design blades can be deduced from this 1D theory.

Therefore Glauert [10] formulated a 2D theory in which circumferential flow is allowed for. So instead of a Cartesian coordinate system (x, y, z) we now use a cylindrical one (x, r, φ) one with the same notations for the velocity components.

To extract power we use now *torque* Q

$$P = F \cdot v = Q \cdot \Omega , \tag{5.14}$$

with Ω being the rotational velocity of the turbine

$$\Omega = \frac{N \cdot \pi}{30} = RPM \cdot 0.1047 . \tag{5.15}$$

From Newton's third law for axial momentum there must be an (angular) momentum balance as well. Therefore, introducing torque and tangential forces, we quantify tangential induced velocities

$$a'(r) \cdot \Omega r, \ x = 0, \tag{5.16}$$

$$2a'(r) \cdot \Omega r, \ x > 0 . \tag{5.17}$$

The total velocity at the disk now consists of two components perpendicular to each other:

$$u_2 = (1 - a)u_1, \tag{5.18}$$

$$u_\varphi = (1 + a') \cdot \Omega r, \tag{5.19}$$

Fig. 5.3 Annuli of extension
dr from $0 \leq r \leq R_{Tip}$

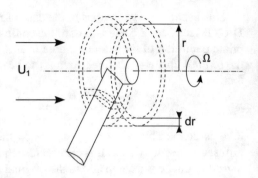

$$in\ total: \ w^2 = u_2^2 + v_t^2. \tag{5.20}$$

We divide the disk into annuli of thickness dr and area $dA = 2\pi r \cdot dr$. (See Fig. 5.3).

The change of angular momentum is

$$dQ := d\dot{m} \cdot v_\varphi \cdot r = \tag{5.21}$$
$$\rho dA u_1 (1-a) \cdot 2a' \Omega r \cdot r\ . \tag{5.22}$$

As $P = Q \cdot \Omega$ also $dP = dQ \cdot \Omega \overset{!}{=} dT \cdot u_2$.

The last equation seems a little bit surprising as one may think that axial and tangential forces should be independent. This is obviously not the case if we use the energy (Bernoulli's) equation across the disk. Specific enthalpy (Bernoulli constant) was already introduced in Sect. 3.2.3, Eq. (3.24). It may be shown[2] [10, 35] that

$$\Delta p = \rho u_\varphi \cdot \left(\Omega r + \frac{1}{2} u_\varphi\right) \tag{5.23}$$

with u_φ from Eq. (5.19). If we now equate

$$dT = 4\pi \rho u_1^2 (1-a) a r dr, \tag{5.24}$$
$$and \tag{5.25}$$
$$dT = \Delta p \cdot 2\pi r dr = 2\pi r u_\varphi \left(\Omega r + \frac{1}{2} u_\varphi\right) r dr, \tag{5.26}$$

we get

$$(1-a)a = \lambda' x^2 (1+a')a' \tag{5.27}$$

with $x := r/R_{Tip}, 0 \leq x \leq 1$ and $\lambda = \Omega R/u_1$ as already defined in Eq. (2.3). The increment per annulus for the power coefficient is

---

[2]Sharpe [8, 32] disagreed with Glauert on how the energy equation for the rotational part is transferred. We come back to this below in Sect. 5.3.

$$dc_p = 8(1 - a)a'\lambda^2 x^3 \, dx \, , \tag{5.28}$$

or if we integrate over all annuli

$$c_p = 8\lambda^2 \int_0^1 (1 - a)a'x^3 \, dx \, , \tag{5.29}$$

To maximize the total power Eq. (5.29), we have to optimize locally Eq. (5.28). For that we define a local efficiency

$$\eta(a, a', x) := a' \, (1 - a) \, x^3 \tag{5.30}$$

and differentiate $\partial\eta/\partial a'$ to obtain

$$\frac{da(a')}{da'} = \frac{1 - a}{a'} \, . \tag{5.31}$$

Doing the same with Eq. (5.27) we have

$$\frac{da(a')}{da'} = \frac{(\lambda \cdot x)^2}{1 - 2a} \, , \tag{5.32}$$

or

$$\frac{1 - a}{a'} = \frac{\lambda^2 x^2}{(1 - 2a)} \, . \tag{5.33}$$

Combining Eqs. (5.27) and (5.33) we may drop $\lambda \cdot x$ to obtain a relation $a'(a)$ only:

$$a' = \frac{1 - 3a}{4a - 1} \, . \tag{5.34}$$

As always $a' > 0$ this implies $1/3 > a > 1/4$ (Fig. 5.4).

To solve for numbers one may start with the cubic equation

$$16a^3 - 24a^2 + 3a(3 - \lambda^2 x^2) - 1 + \lambda^2 x^2 = 0 \tag{5.35}$$

with solution [5] $a = \frac{1}{2}\left(1 - \sqrt{1 + \lambda_r^2}\sin\left[\frac{1}{3}arctan\left(\frac{1}{\lambda_r}\right)\right]\right)$ and $\tag{5.36}$

$$\lambda_r := \lambda \cdot x \tag{5.37}$$

and the rest follows for a' from Eq. (5.34) for the flow angle from

$$tan(\varphi) = \frac{a'}{a} \cdot \lambda x, \tag{5.38}$$

and finally dc_P from Eq. (5.28).

As a remark, it may be noted that Schmitz [31] formulated a comparable model. However, it was published in German only in an unknown journal behind the *iron*

Fig. 5.4 Velocity and force triangles

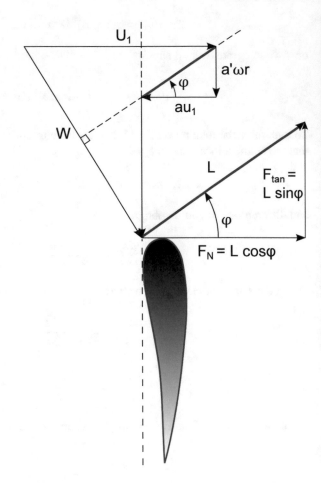

curtain. Nevertheless, a nice asymptotic formula $\lambda \to \infty$ results. To order λ^{-4}:

$$c_P^{Schmitz} = \frac{16}{27}\left(1 - 0.219\frac{1}{\lambda^2} - 0.106\frac{1}{\lambda^4} - \frac{4 \cdot ln(\lambda)}{9 \cdot \lambda^2}\right) . \tag{5.39}$$

It gives $x_P = 0.5852, 0.5703$ and 0.4 for $\lambda = 10, 5$ and 1. These values are in accurate agreement with values from Table 5.1 even for small values of λ (Fig. 5.5).

As a consequence we now may derive a relation for the *chord* of the blade, if we assume that the forces come from lift only

$$\sigma := \frac{Bcc_L}{8\pi \cdot r} = 4Ax\frac{sin^2(\varphi)}{cos(\varphi)} \tag{5.40}$$

$$A := \frac{a}{1 - a} \tag{5.41}$$

Table 5.1 Numerical values for a locally optimized blade according Glauert's general momentum theory

Local TSR	Axial induction	Rotational induction	Flow angle	Power	Solidity
λx	a	a'	φ	c_P	σ
0.25	0.280	1.364	50.6	0.176	0.3658
0.50	0.298	0.543	42.3	0.289	0.5205
1.0	0.316990	0.182981	30.0	0.4154	0.5355
2.0	0.327896	0.052349	17.7	0.5111	0.3799
3.0	0.330748	0.024016	12.2	0.5453	0.2749
4.0	0.331842	0.013669	9.3	0.5615	0.2128
5.0	0.332367	0.008797	7.5	0.5703	0.1728
6.0	0.332658	0.006127	6.3	0.5758	0.1452
7.0	0.332835	0.004510	5.4	0.5795	0.1250
8.0	0.332951	0.003457	4.9	0.5820	0.1098
9.0	0.333031	0.002733	4.7	0.5838	0.0977
10.0	0.333088	0.002215	3.8	0.5852	0.0881
11.0	0.333130	0.001831	3.4	0.5863	0.0802
12.0	0.333163	0.001540	3.2	0.5871	0.0736

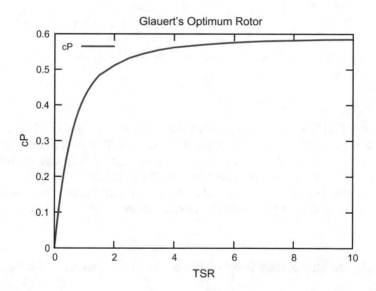

Fig. 5.5 cP versus TSR for a rotor according to Glauert's general momentum theory

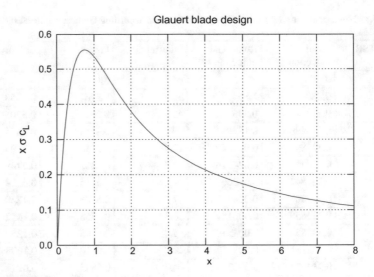

Fig. 5.6 Solidity of a blade from Glauert's theory

Fig. 5.7 Pressure on control
volume

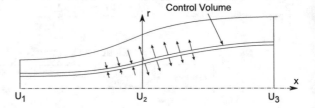

which is shown in Fig. 5.6.

Recently [35] Glauert's model has been reinvestigated especially for low tip speed ratios for the case of constant circulation ([10], p. 193). The original singularity for $\lambda \to 0$ corresponding to $cP \to \infty$ could be eliminated by assuming an influence from the lateral pressure on the curved stream-tube (compare to Fig. 5.7).

A CFD model (see Sect. 7.7 for details) [26, 39] gives a glimpse of how closely (an integral) analytical and (differential) numerical solution may approach each other.

5.3 Limits and Extensions of General Momentum Theories

Momentum theory in itself is a very simple theory. Therefore, is important to know the limits of validity and apply this knowledge to serve as a guide on how to improve this class of models. One important point is the boundary conditions for the (static) pressure far downstream. From a physicist's point of view every quantity should be equally finite (best: 0) for $r \to \infty$. However, some authors argue that the static pressure inside the slipstream might have another value than outside. Clearly, this might be used as an energy source. Sørensen [38] and van Kuik [36, 37] recently have published a more comprehensive discussion. The most general expression which can be deduced is

$$b(1 - a) = \frac{p_3 - p_1}{\rho u_1^2} \frac{1 - a}{1 - b} + \frac{\Delta X}{\rho u_1^2 \Delta A} + 2\lambda^2 x^2 a' (1 + a') \qquad (5.42)$$

in which $b =$ induction in the slipstream far downstream,

$$\Delta X = \oint_{Control\,Volume} p(\mathbf{dA} \cdot \mathbf{e_x}) , \qquad (5.43)$$

the force from the pressure on the slipstream (Fig. 5.8).

With the following assumptions: $b \approx 2a$, $p_0 \approx p_3 \Rightarrow$, and $\Delta X = 0$ Glauert's expressions are recovered:

$$a(1 - a) = \lambda^2 x^2 a'(1 + a') , \qquad (5.44)$$

$$\Delta T = \frac{1}{2}\rho(u_1^2 - u_3^2) \cdot \Delta A , \qquad (5.45)$$

$$u_2 = u_{Disk} = \frac{1}{2}(u_1 + u_3) . \qquad (5.46)$$

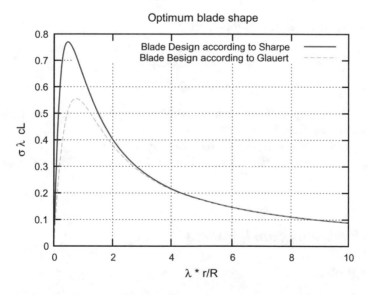

Fig. 5.8 Solidity of a blade from Sharpe's (constant circulation) theory compared with Glauert's

The only way to extend momentum theory is to start with the DEQ approach [39] in Sect. 3.11 and then to try to proceed as far as possible analytically.[3] Unfortunately no—or almost no—purely analytical methods are available, but efficient numerical methods then must help. A group of scientists from Copenhagen have proceeded with the development of this line of theory. A recent summary and discussion of their findings may be found at [36, 37]. As may be expected, the deviations from simple momentum theory become particularly large when the rotor is highly loaded.

In a completely different approach Conway ([7] of Chap. 6) (see Sect. 6.6) has found comparable results.

5.4 The Blade Element Momentum Theory

5.4.1 The Original Formulation

In its simplest formulation, the forces from momentum theory are balanced by forces from airfoils (lift and drag):

$$c_N = c_L cos(\phi) + c_D sin(\phi), \tag{5.47}$$

$$c_{tan} = c_L sin(\phi) - c_D cos(\phi) , \tag{5.48}$$

which then results in equations for a and a':

$$a \quad = \frac{1}{4sin^2(\phi)/(\sigma c_N)+1}, \tag{5.49}$$

$$a' = \frac{1}{4sin(\phi)cos(\phi)/(\sigma c_{tan})-1} . \tag{5.50}$$

Here $\sigma = Bc/2\pi r$ again is the solidity and $N(ormal)$ means the normal while tan the tangential direction.

Clearly this set of equations may be solved by iteration only.

For an ideal rotor one may set $c_L = 2 \cdot \pi$ and $c_D = 0$, together with Eqs. (5.41) and (5.38), respectively.

5.4.2 Engineering Modifications

5.4.2.1 Real Airfoils

Airfoil data in the form of tables of c_L and c_D as a function of angle-of-attack (AOAα) have to be provided. In most cases measurements are not available for AOAs greater

---

[3] van Kuik [17, 18] examined the type of the singularity at the disk-edge in more detail and applied the consequences to real turbines. See also [41].

than α_{stall}. In these cases an empirical correction from Viterna and Corrigan [23] in Chap. 2, for $\alpha < \alpha_{Stall}$ and Aspect Ratio μ:

$$c_L = \frac{c_{D,max}}{2}sin(2 \cdot \alpha) + K_L \cdot \frac{cos^2(\alpha)}{sin(\alpha)}, \tag{5.51}$$

$$c_D = c_{D,max}sin^2(\alpha) + K_D \cdot cos(\alpha), \tag{5.52}$$

$$K_L = (c_{L,S} - c_{D,max}sin(\alpha_S) \cdot cos(\alpha_S)) \frac{sin(\alpha_S)}{cos^2(\alpha_S)}, \tag{5.53}$$

$$K_D = \frac{c_{D,S} - c_{D,max}sin^2(\alpha_S)}{cos(\alpha_S)}, \tag{5.54}$$

$$if \ \mu \le 50: \qquad c_{D,max} = 1.11 + 0.018 \cdot \mu, \tag{5.55}$$

$$if \ \mu > 50: \qquad c_{D,max} = 2.01. \tag{5.56}$$

More about Post-Stall may be found in [43].

Sometimes—especially in the inside parts (low r/R_{tip}) of a blade—the solution of Eq. (5.50) gives values outside $0 \le a \le 1/3$. Then empirical extensions are applied which in their simplest form result in a linear interpolation to an assumed maximum thrust-coefficient $c_{T_{max}} \approx 2$:

$$c_T(a) = 4a(1-a), \ 0 \le a \le 1/3, \tag{5.57}$$

$$c_T(a) = b + m \cdot a, \ 1/3 \le a \le 1, \tag{5.58}$$

and $m = \frac{8}{9} \cdot c_{T,max} - 1, c_{T,max} > 8/9$. Figure 5.9 shows this for $c_{T,max} = 2.0$

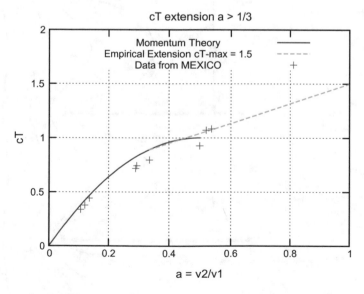

Fig. 5.9 Extrapolation of c_t values for $a > 1/3$ compared to measurements, from [30]

5.4.2.2 Finite Number of Blades

Prandtl [2] showed with an application of potential-theory, Sect. 3.5, that close to the
tip of a blade the circulation must go down to zero, because of pressure compensation
between the upper and lower side at the tip of an airfoil. He compared that situation
to a stack of flat plates, because this model is manageable using elementary methods,
see Fig. 5.10.

Calculating using this model he was able to show the following:

Fig. 5.10 Prandtl's flow
model to derive a simple
model for circulation loss at
the tip

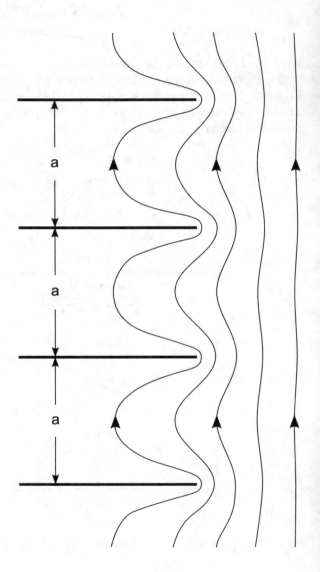

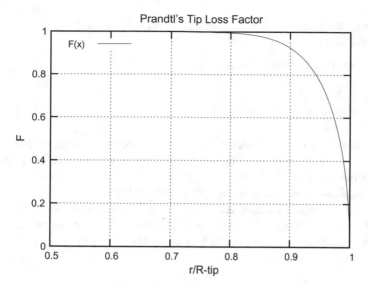

Fig. 5.11 B = 3, a = 0.3, λ(TSR) = 6

$$cos\left(\frac{\pi}{2}F\right) = exp\left(-\frac{B(1 - r/R_{Tip})}{(2r/R_{Tip})sin(\phi)}\right), \qquad (5.59)$$

where F is so called *tip-loss-factor*. Thereby we are able to discuss in approximate terms the effect of finite numbers of blades. It must be stated that still today these *tip-loss models* are discussed because the influence on c_P is rather pronounced. A typical engineering expression states

$$\Delta c_P = const/(B \cdot \lambda), \qquad (5.60)$$

with const ≈ 1.5. Main source of ambiguities is the question of how to apply F to the induction factors, forces and/or airfoil data and how to calculate the flow angle. Ramdin [25] shows that a set of more than 70 combinations are possible. Equation (5.59) appears first in [10] and seems to be the most common implementation. However, Branlard [5] shows that application of some of the proposed tip correction may result in changes of annual energy production of 1 ± 2%. We will come back to this in more detail in Chap. 6. Ambitious readers may solve Problem 5.6. For the most-often used blade number $B = 3$ we have in Fig. 5.13 a variation of airfoil data: If we compare the findings of Fig. 5.13 with Fig. 5.15 we see that a lift-to-drag ratio much larger than 100 is necessary to reach reasonable power production (>0.5 c_P) (Fig. 5.11).

We may close this section by quoting a remark from van Kuik [39] in Chap. 3, p. 104, that there is

ample room for discussion and future work.

5.4.2.3 Behavior Close to the Hub

In addition to the tip region, noticeable changes in aerodynamic forces are also seen in the inboard sections. In case of lift, this was already recognized by Himmelskamp in his Ph.D. thesis [12] in 1950. He used a Gö625 profile with 20% thickness. Compared to the 2D wind-tunnel measurements, he obtained almost a doubled $c_L \sim 3.0$ close to the hub from which he concluded that due to inertial forces separation may be delayed to higher AOAs. Therefore so-called 3D-correction using two parameters (c = chord of blade at radius r) and two constants (a and b) are most commonly written in the form of [36]:

$$c_{L,3D} = c_{L,2D} + a(c/r)^b \left[c_{L,inv} - c_{L,2D} \right] . \tag{5.61}$$

Bak, Johannsen, and Andersen [1] investigated three different rotors by CFD and compared to pure 2D and 3D corrections of other investigations. Since then remarkable progress has been achieved.

On top of that, some authors [8] even argue for *blade root losses* with the same line of argument as for the top. As in most cases $c_l \to 0$ as $r/R_{tip} \to 0$ by use of a cylindrical section as connection to the hub this might be regarded questionable.

5.4.3 Comparison to Actual Designs

As this method is the most widely used in industrial applications for rotor performance and load prediction since it was published 80 years ago, it has to be noted that in general the agreement is not always the best. References [6, 7, 27] give a small insight into how much effort has been spent into validation of these codes (Figs. 5.12 and 5.13).

During the big NASA-Ames blind comparison (see Sect. 8.5, and Ref. [49] there) BEM-Codes have been compared also with so-called *Wake Codes* and CFD (= RANS, see Sect. 7.8). As an example in Fig. 5.14 the comparison for the shaft torque Q ($P = Q \cdot \omega$) is shown. Because the rotor was designed somewhat unconventionally, BEM only gives reasonable results if the stall behavior of the S809 is modeled carefully [43].

In 2006 in Europe's largest wind tunnel, the DNW-LLF, a comparable experiment called *MEXICO* (= Model Experiments in Controlled Conditions) was conducted. Forces, pressures on the blades, and the flow field was measured by PIV [28]. More

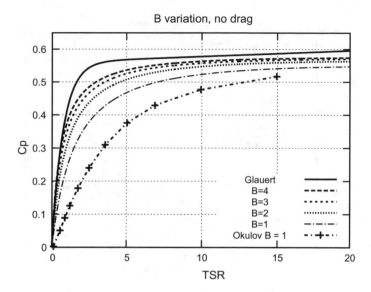

Fig. 5.12 c_P via λ, number of blades B varies

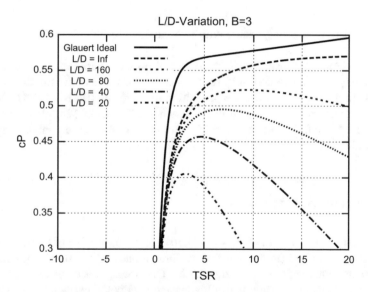

Fig. 5.13 c_P via λ. Number of blades fixed to B $= 3$. Lift-to-drag ratio varies

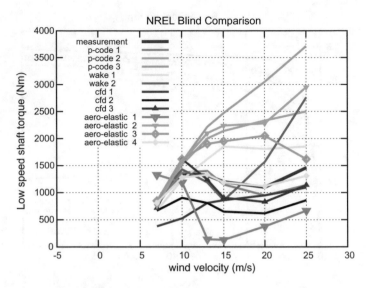

Fig. 5.14 Some results from NASA-Ames blind comparison

about comparing modeling results with experimental results [29] will be presented in Sect. 8.6.

Madsen et al. [21] investigated the effect of pressure variation close to the hub due to wake rotation and non-uniformity of induction close to the tip, thereby considerably improving accuracy of c_P prediction for a high c_P (0.51–0.52) 5 MW rotor.

Much controversy was created in 2004, when ENERCON [24] claimed to have measured $c_P^{max} = 0.53$ (E2 of Fig. 5.15). Somewhat later these peak-values were reduced to 0.51.

5.5 Optimum Rotors I

For a long time many authors [23, 44, 45], to mention only a few of them, have tried to design an optimized (in the sense of reaching highest c_P^{max}) rotor which includes the presented findings of engineering BEM. One weak spot is, of course, that the model assumes an infinite number of blades and the Prandtl correction, Eq. (5.59) is valid only to a nominal accuracy.

It is, therefore, necessary to use a different approach, the vortex model of wind turbines, to look for an *optimum* rotor. This will be presented in the next chapter.

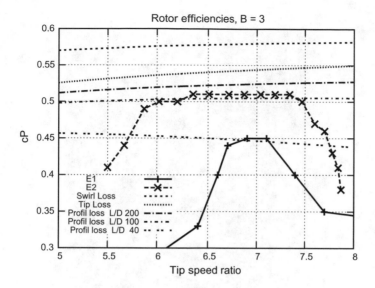

Fig. 5.15 Comparison of efficiency of 3-bladed wind turbines

Fig. 5.16 Double actuator disk as a BEM model for vertical axis wind turbines

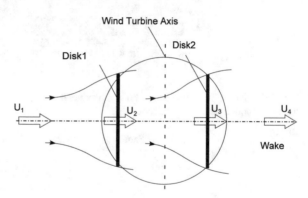

5.6 Vertical Rotors

BEM may be adapted to Vertical Axis Rotors as well. As the motion is instationary and encompasses a whole volume (something between a cylinder and a sphere) the simplest possible model consists of two disks, see Fig. 5.16, and [20] in Chap. 2. It was discussed in detail by [8] in Chap. 2 using vortex methods, see Sect. 6.3. In Figs. 5.18 and 5.19 we see a comparison of performance to measurements [2] in Chap. 2 for a 34 m diameter 2-bladed *egg-beater*, see Fig. 5.17.

Fig. 5.17 Sandia 34 m diameter turbine, reproduced with permission of Sandia National Laboratories, Albuquerque, New Mexico, USA

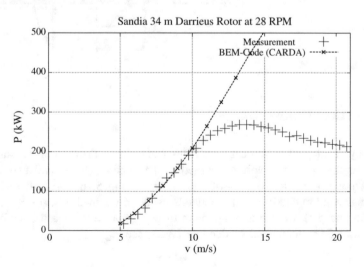

Fig. 5.18 Sandia 34 m diameter turbine: power as function of wind speed

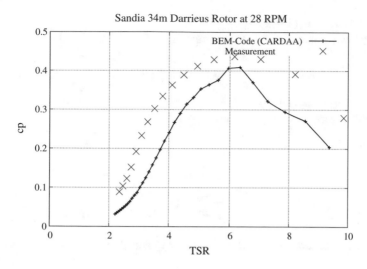

Fig. 5.19 Sandia 34 m diameter turbine: c_p as function of TSR

5.7 Problems

Problem 5.1 Strange Farmer?[4]

(a) A wind turbine with $D = 33$ m and a rated power of $P = 330$ kW may serve as a fan. Show by simple momentum theory:

$$T = \sqrt[3]{2\rho A_r \cdot P^2} \tag{5.62}$$

$$v_2 = P/T \tag{5.63}$$

and finally estimate the maximum possible blowing velocity $v_3 == 2 \cdot v_2$.

(b) Now make a more realistic BEM model by using reasonable blade geometry data and Cl-Cd-data and vary pitch until either maximum **output** power or pitching angle is reached. Make sure that your BEM model may switch properly from turbine $(a > 0)$ to fan mode $(a < 0)$.

Problem 5.2 Multiple Actuator Disks

A Darrieus-type VAWT serves as a simple model of a counter-rotating (see Sect. 2.7) *double actuator disk*. The model assumes two fully developed slipstream, and the far downstream outlet of the first disk is regarded as the inlet of the second one. So we have five special locations and two induction factors a and b:

- 1: far upstream whole assembly, u_1.

---

[4]In fact this problem was real. In 1993 a farmer from the German West Coast (close to the North Sea) asked if an ENERCON33 wind turbine could serve as a fan to blow warmer air from above to his blooming apple trees to protect them against night frost.

- 2: disk 1, $u_2 = u_1(1 - a)$.
- 3: far downstream disk 1 = far upstream disk 2,

$$u_3 = u_1(1 - 2a). \tag{5.64}$$

- 4: disk 2, $u_4 = \mathbf{u_3}(1 - b)$.
- 5: far downstream whole assembly, $u_5 u_3(1 - 2b)$

(a) show that

$$c_{P1} = 4a(1 - a)^2, \tag{5.65}$$

$$c_{P2} = \frac{u_3^3}{u_1^3} 4b(1 - b)^2. \tag{5.66}$$

(b) maximize total power $c_P^{total} = c_{P1} + c_{P2}$ and show

$$b_{max} = \frac{1}{3}, \tag{5.67}$$

$$c_{P2} = 4a(1 - a)^2 + (1 - 2a)^3 \cdot \frac{16}{27} \rightarrow max \tag{5.68}$$

$$a_{max} = 0.2(!), \tag{5.69}$$

$$c_P^{max,total} = \frac{64}{125} + \frac{16}{125} = 0.64 > 0.59. \tag{5.70}$$

** (c) Show that for $N \in \mathbb{N}$ disks:

$$c_p^{max,Ndisks} = \frac{8}{3} \cdot \frac{1 + (1/N)}{(2 + (1/N)^2)}, \tag{5.71}$$

$$(N \rightarrow \infty) \rightarrow \frac{2}{3} = 0.67 . \tag{5.72}$$

Problem 5.3 Use the code from Problem 5.1 to deduce the power curve NTK for 500/41 from [26] in Chap. 3.

Problem 5.4 ([23] in Chap. 2) uses a kind of a Rankine vortex-model (see Sect. 3.6, Fig. 3.11) to include wake rotation. Try to show and understand them in finding the following:

$$c_P = \frac{b^2(1 - a)^2}{b - a} \tag{5.73}$$

$$= \frac{b^2(1 - a)^2}{b - a} [b + (2a - b)\Omega/\omega_{max}] . \tag{5.74}$$

Here ω_{max} is the maximum velocity at the edge of the core.

Problem 5.5 BEM model of a wind-driven vehicle using a turbine:

If a wind turbine is used to drive a vehicle (with velocity w), thrust has to be transported, the extracted power now being

$$P = (T + D)w, \tag{5.75}$$

$$\text{with } D = \frac{1}{2}\rho A_{vehicle}(u_1 + w)c_D . \tag{5.76}$$

(a) Estimate Force and Power for a wind car with $4\,\text{m}^2$ rotor-area. Assume $v_{wind} = 8\,\text{m/s}$ and $v_{car} = 4\,\text{m/s}$.

(b) Show [40] that for maximum car-speed $w/u_1 \to max$ the (usually defined) induction a depends only on one parameter K:

$$a^3 - a^2 - 3Ka + K = 0 , \tag{5.77}$$

with

$$K := \frac{1}{4}\left(\frac{A_{vehicle}}{A_{rotor}}\right) \cdot c_d . \tag{5.78}$$

(c) Solve this equation for a(K) and prepare a table with columns:
a, w, c_P, c_T and c_P/c_T of K with $0.001 \le K \le 10$.
(d) Prepare a (double-logarithmic) graph and discuss the results.
** (e) now include drive train efficiencies [9].

Problem 5.6 Show by conformal mapping how Eq. (5.59) may be deduced.

Problem 5.7 Show by using a BEM code and empirical equations for ideal chord, twist and lift

$$c(r/R_{Tip}) = \frac{16\pi}{9}\left((\lambda r/R_{Tip})^2 + 4/9\right)^{-1/2} \tag{5.79}$$

$$tan(\theta) = \frac{1}{\lambda \cdot r/R_{Tip}} \tag{5.80}$$

$$c_L = 2\pi\alpha \tag{5.81}$$

how Figs. 5.12 and 5.13 may be deduced.

References

1. Bak C, Johansen J, Andersen PB (2006) 3D-corrections of airfoil characteristics based on pressure distributions. In: Proceedings of the EWEC, Athens, Greece
2. Betz A (1919) Schraubenpropeller mit geringstem Energieverlust. mit einem Zusatz von L. Prandtl. Nachr d Königl Gesell Wiss zu Göttingen, Math-phys Kl 198–217
3. Betz A (2013) Das Maximum der theoretisch möglichen Ausnützung des Windes durch Windmotoren, Z ges Turbinenwesen 26:307–309, English translation: (1920) Wind engineering, 32, 4

4. Betz A (1926) Wind-Energie und ihre Ausnutzung durch Windmühlen. Vandenhoeck und Ruprecht, Göttingen
5. Branlard E (2017) Wind turbine aerodynamics and vorticity-based methods. Springer International Publishing, Berlin
6. Buhl ML, Wright AD, Tangler JL (1997) Wind turbine design codes: a preliminary comparison of the aerodynamics, NREL/CP-500-23975, NREL, CO, USA
7. Buhl ML, Manjock A (2006) A comparison of wind turbine aero elastic codes used for certification, NREL/CP-500-39113, NREL, CO, USA
8. Burton T, Sharpe D, Jenkins N, Bossanyi E (2011) Wind energy handbook, 2nd edn. Wiley, Chichester
9. Gaunå M, Øye S, Mikkelsen R (2009) Theory and design of flow driven vehicles using rotors for energy conversion. In: Proceedings of the EWEC, Marseille, France
10. Glauert H (1936) Aerodynamics of propellers. J. Springer, Berlin
11. Hansen MOL (2004) Aerodynamics of wind turbines, 2nd edn. Earthscan, London
12. Himmelskamp H (1950) Profiluntersuchungen an einem umlaufenden Propeller, Mitt Max-Planck-Inst f Strömungsforschung, vol 2, Göttingen, Germany (in German)
13. Hansen MOL, Johansen J (2004) Tip studies using CFD and comparison with tip loss models. In: Proceedings of the 1st conference on the science of making torque from wind, Delft, The Netherlands
14. Jamieson P (2008) Generalized limits for energy extraction in a linear constant velocity flow field. Wind Energy 11(5):445–457
15. Jamieson P (2008) Private communication
16. Jamieson P (2018) Innovation in wind energy, 2nd edn. Wiley, Chichester
17. van Kuik GAM (1991) On the revision of the actuator disc momentum theory. Wind Eng 15:276–289
18. van Kuik GAM (2003) The edge singularity of an actuator disk with a constant normal load, AIAA paper 2003-356, Reno NV, USA
19. van Kuik GAM (2018) The fluid dynamic basis for actuator disc and rotor theories. IOS Press, TU Delft, The Netherlands
20. Loth JL, McCoy H (1983) Optimization of Darrieus turbines with an upwind and downwind momentum model. J Energy 7:313–318
21. Madsen HA et al (2010) Validation and modification of the blade element momentum theory based on comparisons with actuator disc simulations. Wind Energy 13:373–389
22. Munk M (1920) Über vom Wind getriebene Luftschrauben. Z f Flugtechnik u Motorluftschiff XI(15):200–223 (in German)
23. Mikkelsen R, Øye S, Sørensen JN (2006) Towards the optimally loaded actuator disc. In: Proceedings of the IAEwind annex XI (basic information exchange), Kiel, Germany
24. NN (2004) Considerably higher yields - revolutionary rotor blade design. Wind Blatt 3, Aurich, Germany
25. Ramdin SF (2017) Prandtl tip loss factor assessed, MSc thesis, TU Delft, Delft, The Netherlands
26. Schaffarczyk AP (2012) AD-CFD implementation of the Glauert-rotor, Unpublished internal report, 89, UAS Kiel, Germany
27. Schepers JG (2002) Verification of European wind turbine design codes, VEWTDC, Final report, ECN-C-01-055, ECN, The Netherlands
28. Schepers JG, Snel H (2007) Model experiments in controlled conditions, Final report, ECN-E-07-042, ECN, The Netherlands
29. Schepers JG et al (2012) Analysis of Mexico wind tunnel measurements, Final report of IEA task 29, Mexnext (phase 1): ECN-E-01-12-004. ECN, The Netherlands
30. Schepers JG, Boorsma K, Munduate X (2012) Final results from Mexnext-I: analysis of detailed aerodynamic measurements on a 4.5 diameter rotor places in the large German Dutch wind tunnel NNW. In: Proceedings of the 4th conference of the science of making torque from wind, Oldenburg, Germany
31. Schmitz G (1955) Therie und Entwurf von Windrädern optimaler Leistung. Wiss Zeitschr Uni Rostock, 1955/1956 (in German)

32. Sharpe DJ (2004) A general momentum theory applied to an energy-extracting actuator disk. Wind Energy 7:177–188
33. Shen WZ, Mikkelsen R, Sørensen JN, Bak C (2003) Validation of tip corrections for wind turbine computations. In: Proceedings of the EWEC, Madrid, Spain
34. Shen WZ, Mikkelsen R, Sørensen JN, Bak C (2005) Tip loss corrections for wind turbine computations. Wind Energy 8:457–475
35. Sørensen JN, van Kuik GAM (2010) General momentum theory for wind turbines at low tip speed ratios. Wind Energy 43:427–448
36. Sørensen JN (2011) Aerodynamic aspects of wind energy conversion. Annu Rev Fluid Mech 43:427–448
37. Sørensen JN, Mikkelsen R (2012) A critical view on the momentum theory. In: Proceedings of the 4th conference on the science of making torque from wind, Oldenburg, Germany
38. Sørensen JN (2016) General momentum theory for horizontal axis wind turbines. Springer, Cham
39. Sørensen JN (2012) Private communication
40. Sørensen JN, Aero-mekanisk model for vindmølledrevet køretøj, Copenhagen, Denmark, no year
41. Spalart P (2003) On the simple actuator disk. J Fluid Mech 494:399–405
42. Spera DA (ed) (2009) Wind turbine technology, 2nd edn. ASME Press, New York
43. Tangler JL, Kocurek JD (2005) Wind turbine post-stall airfoil performance characteristics guidelines for blade-element momentum methods, NREL/CP-500-36900 and 43rd AIAA aerospace meeting and exhibition, Reno, Nevada, USA
44. de Vries O (1979) Fluid dynamic aspects of wind energy conversion. AGARDograph, vol 242. Neuilly sur Siene, France
45. Wilson RE, Lissaman PBS, Walker SN (1976) Aerodynamic performance of wind turbines. Oregon State University, Corvallis

Chapter 6
Application of Vortex Theory

6.1 Vortices in Wind Turbine Flow

It was already mentioned in Chap. 3 that Saffman [27] quoted Küchemann, who said that

> vortices are the sinews and muscles of fluid motion.

It has to be noted, that based on some progress in the 2D case, some authors [2] even argue for a vorticity-based theory of turbulence.

As shown in Chap. 3 this metaphor can be expressed with equations. Remember that local vortex strength, or vorticity was defined via Eq. (3.39)

$$\omega := \nabla \times \mathbf{v} .$$

Now if a velocity field $\mathbf{v}$ satisfies $\nabla \cdot \mathbf{v} = 0$ (is *solenoidal*), then

$$\mathbf{u} = \nabla \times \mathbf{A} .$$

with $\mathbf{A}$ the vector potential Eq. (3.44). Now ω being given, $\mathbf{A}$ and therefore $\mathbf{u}$ is given by *Poisson's equation* (inhomogeneous Laplace equation):

$$\nabla(\nabla \cdot \mathbf{A}) - \nabla^2 \mathbf{A} = \omega \qquad (6.1)$$

Now, if a flow has (large) regions of vanishing vorticity, we may introduce the concept of *vortex lines* and *vortex sheets*. As can be immediately seen, this is in conflict with the conventional smoothness assumption for the flow and vorticity fields.

© Springer Nature Switzerland AG 2020
A. P. Schaffarczyk, *Introduction to Wind Turbine Aerodynamics*,
Green Energy and Technology, https://doi.org/10.1007/978-3-030-41028-5_6

This is the place where δ functions[1] are introduced. We refer the interested reader to [24] of Chap. 3 for more mathematical rigor. In the simplest case for a vortex line we may write formally [27]:

$$\omega = \Gamma \delta(n)\, \delta(b)\, \mathbf{t}, \tag{6.3}$$

where Γ (Eq. (3.40)) was the circulation and $\mathbf{t}$, $\mathbf{n}$, and $\mathbf{b}$ are the tangent, normal, and b-normal of the space curve. A closed loop is called *vortex ring*. In case of an irregular shape, we talk about *vortex filaments*.

The properties of vortex elements are governed by **Helmholtz' Theorem**, which will be presented in a more modern nomenclature:

Theorem 6.1 *In inviscid flow, in which only conservative forces act*[2]

- *fluid particles free of vorticity remain so,*
- *vortex lines move with the fluid,*
- *circulation is conserved over time.*

See [33] for a rigorous derivation of these conservation laws and a possible connection to turbulence, even in the inviscid case.

To summarize: ordered by dimension, the following forms of sustained vorticity ay be important to wind turbine flow and will be discussed in more detail in the text that follows:

P   0D patches: 2D transient VAWT,
L   1D lines of finite length: Diffuser design,
H   1D helical lines of infinite length: Joukovski,
S   2D vortex sheets: Betz/Goldstein Rotor,
V   3D vortical wake volume: Conway.

But before we start, we should introduce some well-known quantifiable examples.

6.2   Analytical Examples

As an initial example, we now generalize from Sect. 3.6, Fig. 3.11, the *Rankine vortex* which has a non-smooth slope at the core's border. The so-called $n = 2$ *Scully vortex* (see [6] of Chap. 1)

---

[1] In fact this is a distribution (sometimes also called a generalized function) with the property that

$$\forall f \in \mathscr{S} : \int \delta(x) f(x)\, dx = f(0) \tag{6.2}$$

$f(x)$ is called a *test function*. $\mathscr{S}$ is the set of test functions named after *Laurent Schwartz*, Field's medal winner of 1950.

[2] Von Kuik [15], [19] in Chap. 5 recently pointed out how to combine non-conservative forces with the conventional lifting-line Helmholtz–Kelvin approach.

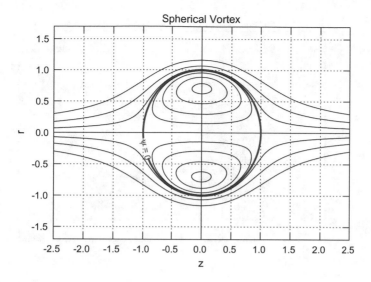

Fig. 6.1 Hill's spherical vortex

$$v_\phi(r) = \left(\frac{\Gamma}{2\pi r_c} \frac{r}{\sqrt[n]{1 + r^{2n}}} \right) \tag{6.4}$$

gives a somewhat smoother shape. Reference [16] gives pictures of recent measurements of *tip-vortices* in water. During the MexNext experiment, the shape and movement of the vortex was measured in some detail ([29, 30], Chap. 5).

A well-known (second) example is *Hill's Spherical Vortex* with vorticity in cylindrical coordinates $\omega = (\omega_r = 0, \omega_\phi, \omega_z = 0)$ (Fig. 6.1):

$$\omega_\phi = \begin{cases} A \cdot r & \text{if } r^2 + z^2 < a^2 , \\ 0 & \text{if } r^2 + z^2 > a^2 . \end{cases} \tag{6.5}$$

The stream function is

$$\Psi = \begin{cases} \frac{-A}{10} \left(r^4 + r^2 z^2 + \frac{5}{3} r^2 a^2 \right) & \text{if } r^2 + z^2 < a^2 , \\ \frac{A r^2 a^5}{15(r^2 + z^2)^{3/2}} & \text{if } r^2 + z^2 > a^2 . \end{cases} \tag{6.6}$$

It has the remarkable property, that it moves with a self-induced velocity of

$$u = 2Aa^2/15 \tag{6.7}$$

through the fluid.

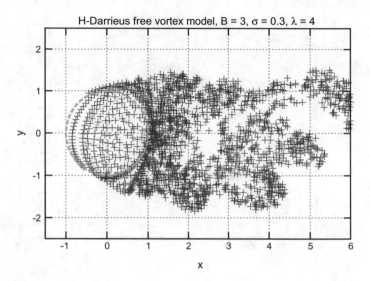

Fig. 6.2 Shedded vorticity patches

6.3   Vortex Patches

Vortex patches are small finite volumes (in 3D) or areas (in 2D) with a finite amount
of vorticity. A very simple and easy to calculate example is the H-shaped VAWT
where due to conservation of circulation, vorticity is continuously shedded (compare Fig. 6.2), because AOA and therefore lift changes. The motion of the patches
is calculated by Helmholtz's second law. As an example, a 2D H-Darrieus (from
Chap. 2, [8]) was coded (in FORTRAN, see listing in Problem 6.5) with the method
from [32]. As a result, we see in Fig. 6.3 the evolution of the normal and tangential
forces as well as the power coefficient.

6.4   Vortex Lines

6.4.1   *Vortex Lines of Finite Length*

Extending the almost point-wise *patches* to a line of finite length gives a good example
of investigations of the action of a diffuser (see Sect. 2.5 and Fig. 2.11) if regarded as
a *ring wing*. We follow Appendix A of [44]. If we associate the bound vorticity Γ_{bound}
with a simple circular ring vortex of zero core-size, we may find [27] arbitrary large
values close to the ring. de Vries found an elegant way out—apparently inspired by
thin-airfoil theory—and assumed distributed $(0 \leq x \leq c)$ vorticity along the whole
chord. After some algebra on elliptic integrals [1]:

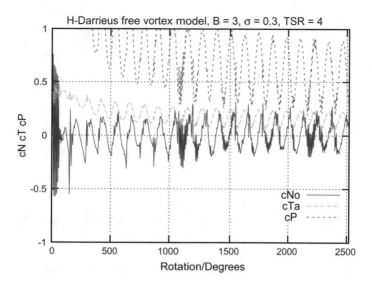

Fig. 6.3 Evolution of forces after seven revolutions

$$E(k) := \int_0^{\pi/2} d\phi \sqrt{1 - k^2 \, sin^2(\phi)}, \tag{6.8}$$

$$K(k) := \int_0^{\pi/2} \frac{d\phi}{\sqrt{1 - k^2 \, sin^2(\phi)}}, \tag{6.9}$$

he was able to provide an approximate expression for the averaged induced velocity, see Fig. 6.4. This concept was successfully applied to diffuser designs for small wind turbines up to 5 kW by Schaffarczyk [29].

6.4.2   Preliminary Observations on Vortex Core Sizes

As already noted, it may be shown easily that a straight vortex line with concentrated vorticity $\omega = \Gamma \delta(x)$ induces a velocity field with specific energy (J/m) of

$$E \sim \log(R/r_{core}). \tag{6.10}$$

which is singular for $r_{core} \rightarrow 0$ and $R \rightarrow \infty$ as well.

Fortunately, due to viscosity we always have a finite core sizes. For example [39] provides for tip-vortices of helicopter blades an empirical formula

$$ln\left(r_{core}/R_{tip}\right) \sim c_T^{-0.5}. \tag{6.11}$$

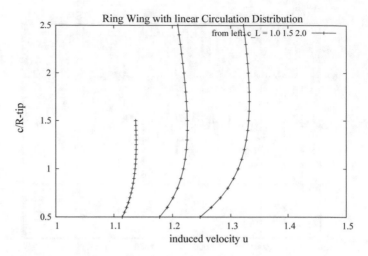

Fig. 6.4 Averaged induced velocity of a ring wing with diameter R_{tip}, chord c, and lift coefficient c_L

Note that for a propeller c_T is defined as

$$c_T = T/\rho A_R (\omega \cdot R_{tip})^2 \ . \tag{6.12}$$

During the European MEXICO experiment (see Sect. 8.6) velocity fields were measured by *PIV*, so that the tip-vortices could be observed. Figure 6.5 gives some results [29] from Chap. 5.

6.4.3 Helical Vortex Lines

From early flow visualization for ship propellers (Chap. 1, [3], Joukovski [14]) (see Fig. 6.6) deduced a very tidy and simple picture in which the **constant** bound circulation Γ of each blade is convected on helical lines by the flow. On the axis of rotation they combine to the so-called *hub vortex*. Unfortunately this picture could not be used for numerical predictions due to calculation complications because of the singular behavior close to the vortex line.

In a series of papers, Okulov [10, 18, 20–22] used a new method to develop a consistent description. He introduced the concept of *induced equilibrium motion of a multiple of helical vortices* to determine a vortex core of size ε of constant vorticity. The consistency condition [22] (in a slightly adapted notation) reads

$$\bar{u}_\chi \left(\varepsilon/R_{tip}\right) = -\frac{B^2}{\tau}\left(1 + \frac{\varepsilon}{R_{tip}}\right) \tag{6.13}$$

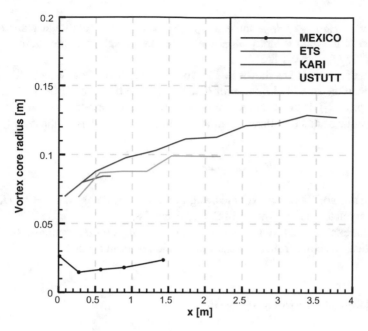

Fig. 6.5 Core radius of the MEXICO-Rotor-Vortex at 10 m/s inflow condition. With permission of University of Stuttgart, Germany

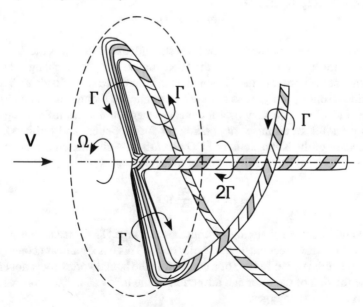

Fig. 6.6 Vortex system from Joukovski [14] for a two-bladed propeller

with $\tau = tan(\varphi) = \ell/R_{tip}$, and φ the angle between the rotational plane and the direction of the total velocity V_0 at the blade. As a result this core radius is on the order of $B\varepsilon/R_{tip} \approx 0.4$ [22].

Numerical values are shown in Table 6.1 and indicate a flow state of higher efficiency than the so-called *Betz rotor* (Chap. 5, [2]). It has to be noted that Burton et al. (Chap. 5, [8]) deduced a constant circulation by other arguments.

The value of this optimized circulation can be expressed as (Chap. 5, [8]):

$$\Gamma_{opt,constant} = 4\pi \frac{u_1^2}{\omega} \cdot \frac{2}{9} . \tag{6.14}$$

From this equation, very simple design rules may be deduced ($\sigma = Bc/R_{tip}$, the shape parameter) as was shown in Figs. 5.6 and 5.8.

Recent experiments by Leweke et al. [16] investigated the dynamic stability of the vortical structure from a constant-circulation rotor in a water tunnel.

6.5   Helical Vortex Sheets

Vortex sheets were introduced in Sect. 3.6.4. As for the vortex line, Eq. (6.3), we may define formally (Fig. 6.7):

$$\omega = \kappa\delta(n), \tag{6.15}$$

where n is the local normal and κ is the local strength of the sheet. As the flow normal to that sheet must be constant, only the tangential velocity may change. This is in fact the case at the border of the 1D-slipstream, where we have outside u_1 but inside all velocities from u_1 to u_3. Also a vortex sheet may be thought of as an alignment of vortex lines. This is a very useful concept when we discuss finite wings either translating (as for an airplane) or rotating (as for propellers and turbines). In any case—because of the Kutta–Joukovski Theorem 3.55—circulation can be related to the chord c and local lift coefficient c_L of a wing:

$$\Gamma = \frac{c_L}{2} w c. \tag{6.16}$$

Now if c and/or c_L are changing, Γ must also change. That means that in case of $d\Gamma/dr \neq 0$ a vortex sheet must necessarily occur. This has significant consequences on the shape of the wake: In case of a constant Γ the bound vortex must not end, and due to the rotation of the rotor and advection by the inflow, a helical vortex line like in Fig. 6.6 from [14] emerges.

This is in sharp contrast to the case of varying circulation. Inspired by Prandtl's *Tragflügeltheorie, (lifting-line theory)* [24, 25],[3] Betz investigated the problem of

---

[3]The most famous outcome was a quantitative description of the phenomenon of induced drag $c_{D,ind} = c_L^2/(\pi AR)$ which is minimized by an elliptical distribution of lift only. Corten [8] shows

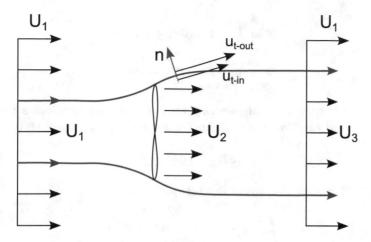

Fig. 6.7 Example of emergence of a simple vortex sheet (seen as jump in tangential velocity) for an actuator disk with constant loading

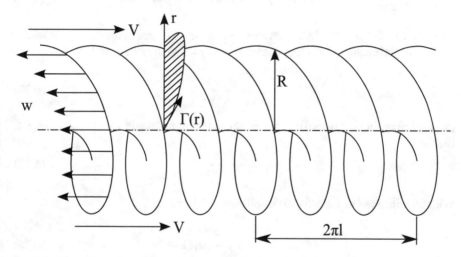

Fig. 6.8 Betz wake

how to design a ship propeller with given thrust and minimal power. He found that the wake—a vortex sheet, see Fig. 6.8—must be of constant *pitch*.

We follow Chap. 1, [3] for a reliable derivation. (Compare to [38] and [5], Chap. 12 for inclusion of drag—both from Sect. 5.7) From the Kutta–Joukovski Theorem, we have for Thrust T and Torque M (density of fluid $\rho = 1$):

---

similarities and differences in the case of wind turbines. He found an AOA correction of $c_L/(8\pi) \cdot (Bc/r) \cdot \lambda_r$.

$$T = \int_0^{R_{tip}} \Gamma \cdot \omega r (1 + a') dr \; , \tag{6.17}$$

$$M = u_1 \cdot \int_0^{R_{tip}} \Gamma (1 - a) r \, dr \tag{6.18}$$

As we want to be $P = \omega \cdot M \to min$ with $T = const$ we introduce a *Lagrange Multiplier* ℓ and use *calculus of variations* that for infinitely small amounts (variation) δT and δM

$$\delta T - \ell \cdot \delta M = 0 \tag{6.19}$$

must hold. Taking the variation of Eq. (6.18)

$$\delta T = \int_0^{R_{tip}} \delta \Gamma \cdot \omega r (1 + a') - \Gamma \frac{\partial a'}{\partial \Gamma} \delta \Gamma dr \; , \tag{6.20}$$

$$\delta M = u_1 \cdot \int_0^{R_{tip}} \delta \Gamma (1 - a) - \Gamma \frac{\partial a}{\partial \Gamma} \delta \Gamma r \, dr. \tag{6.21}$$

To make Eq. (6.21) solvable we have to connect the induced velocities to the circulation. For the tangential component we have by Biot–Savart's Law [5]:

$$a' = \frac{\Gamma(r)}{4\pi \omega} \tag{6.22}$$

(note the extra factor 2 in the denominator!) and (compare again with [5] and Chap. 1, [3]):

$$a = \frac{\Gamma(r)\omega}{4\pi u_1^2} \; . \tag{6.23}$$

With both inserted into Eq. (6.19) it follows:

$$\int_0^{R_{Tip}} \delta \Gamma \underbrace{\left(\omega r (1 + 2a') - \ell u_1 (1 - 2a) r \right)}_{(*)} dr \tag{6.24}$$

Now Eq. (6.24) may only hold for arbitrary $\delta \Gamma$ if

$$(*) = \omega r (1 + 2a') - \ell \cdot u_1 (1 - 2a) r \tag{6.25}$$

vanishes, so that

$$\ell = \frac{\omega}{u_1} \frac{1 + 2a'}{1 - 2a} \tag{6.26}$$

must hold independent from r.

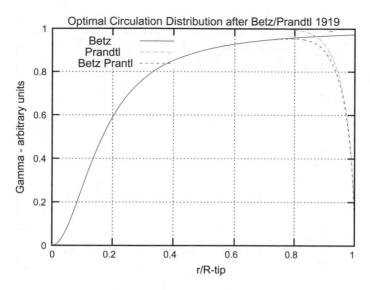

Fig. 6.9 Betz' optimum circulation distribution for a propeller, $B = 3$, TSR $= 6$

Putting Eqs. (6.22) and (6.23) into Eq. (6.26) we arrive at:

$$\Gamma_{Betz,opt} = 2\pi \, (\omega - \ell u_1) \, \frac{r^2}{1 + \ell \cdot \left(\frac{\omega}{u_1}\right) r^2} \tag{6.27}$$

Betz (Chap. 5, [2]) conceived this in a slightly different manner. Figure 6.9 shows the original Betz circulation distribution together with the empirical tip correction in Eq. (5.59) which was given by Prandtl in the same paper as an appendix. It has to be added that—like in momentum theory—it is implicitly assumed:

$$a(x = x_{disk}, r) = \frac{1}{2} a(x \to \infty, r) \text{ and} \tag{6.28}$$

$$a'(x = x_{disk}, r) = \frac{1}{2} a'(x \to \infty, r) . \tag{6.29}$$

Betz' derivation is valid only for $B \to \infty$ and $c_T \to 0$ and shows no decrease in circulation close to the tip. When using Prandtl's simple correction, Eq. (5.59), further approximations are introduced, creating a need for a more precise derivation from first principles.

This was done first by Goldstein [11] (still valid for the case of lightly loaded propeller) and further improved by Theodorsen [34] for moderate loading.

His solution for the velocity potential was given as an infinite series of *Bessel functions*, see [1]. He was able to give figures for two- and four-bladed propellers only. With the advent of digital computers in the 1950s, more accurate and faster calculations were possible, see Tibery and Wrench [35].

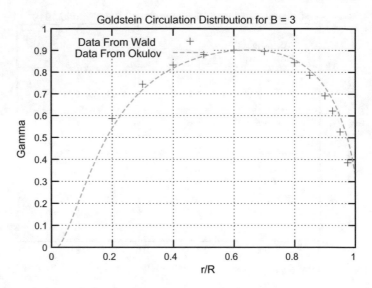

Fig. 6.10 Goldstein's optimum circulation, B $= 3$, $\lambda = 6$. Data from [19, 21, 36]

Okulov [21], based on his solution [10, 18] for the stability problem of flow of helical vortices, improved the numerical schema further. Figure 6.10 gives an example distribution for B $= 3$ and TSR $= 6$. The connection of $G(x, \ell)$ to power is given by [21]

$$c_P = 2\bar{w}\left(1 - \frac{1}{2}\bar{w}\right)\left(I_1 - \frac{1}{2}I_3\right) \tag{6.30}$$

with

$$I_1 = 2\int_0^1 G(x, \ell)x \, dx, \tag{6.31}$$

$$I_3 = 2\int_0^1 G(x, \ell)\frac{x^3}{x^2 + \ell^2} \, dx . \tag{6.32}$$

The connection between ℓ, $\bar{w}$ and our common tip speed ratio $\lambda = \omega r_{tip}/u_1$ is given by

$$\ell = \frac{1 - \bar{w}}{\lambda} \tag{6.33}$$

and $\bar{w}$ the circumferential averaged translation speed of the sheet (see Fig. 6.7).

In the limit $B \to \infty$ Okulov is able to show [21]:

$$G_\infty = \frac{x^2}{x^2 + \ell^2} . \tag{6.34}$$

6.6 Stream Function Vorticity Formulation

6.6.1 Nonlinear Actuator Disk Theory

In a series of papers Conway [4–6], based on [37] formulated a semi-analytical theory for actuator disks ($B \to \infty$) with vortex rings $\omega_{\varphi(r,z)}$ as the basic objects. The slipstream (see Fig. 6.11) is thought to be consisting of **azimuthal vorticity** this being equivalent to a *volume distribution of ring vortices*.

In cylindrical coordinates, the axisymmetric velocity fields induced at a general observation point (r, z) by a single ring vortex of radius r' and strength Γ placed at axial position z' can be constructed by superposition from the elementary solutions of Laplace's equation. The results are

$$\Psi(r, z) = \frac{\Gamma r r'}{2} \int_0^\infty J_1(sr') J_1(sr) e^{-s|z-z'|} ds, \qquad (6.35)$$

$$V_r(r, z) = \frac{-\Gamma \, sign(z - z')r'}{2} \int_0^\infty s J_1(sr') J_1(sr) e^{-s|z-z'|} ds, \qquad (6.36)$$

$$V_z(r, z) = \frac{\Gamma r'}{2} \int_0^\infty s J_1(sr') J_0(sr) e^{-s|z-z'|} ds. \qquad (6.37)$$

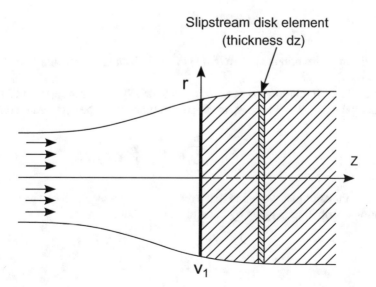

Fig. 6.11 Conways picture of an actuator disk

By superposition, the flow fields induced by the azimuthal vorticity distribution $\omega_\phi(r, z)$ in the slipstream are given by

$$\Psi(r, z) = \frac{r}{2} \int\limits_0^\infty \int\limits_0^{R(z)} \int\limits_0^\infty \omega_\phi(r', z')r'J_1(sr')J_1(sr)e^{-s|z-z'|}ds\,dr'\,dz', \qquad (6.38)$$

$$V_r(r, z) = \frac{1}{2} \int\limits_0^\infty \int\limits_0^{R(z)} \int\limits_0^\infty \pm\omega_\phi(r', z')r's\,J_1(sr')J_1(sr)e^{-s|z-z'|}ds\,dr'\,dz', \qquad (6.39)$$

$$V_z(r, z) = \frac{1}{2} \int\limits_0^\infty \int\limits_0^{R(z)} \int\limits_0^\infty \omega_\phi(r', z')r's\,J_1(sr')J_0(sr)e^{-s|z-z'|}ds\,dr'\,dz'. \qquad (6.40)$$

The vorticity within the slipstream is related to the specific enthalpy h, which is constant along streamlines, by the equation

$$\frac{\omega_\phi(r, z)}{r} = \left(\frac{h - h_0}{(\Omega r)^2} - 1\right)\frac{dh}{d\Psi}, \qquad (6.41)$$

where Ω is the angular velocity of the propeller and h_0 the free-stream specific enthalpy. Specifying the slipstream enthalpy as a function of Ψ defines the circulation distribution and hence the load distribution along the propeller blades. For the case of a counter-rotating propeller, Eq. (6.41) is replaced by the simpler equation:

$$\frac{\omega_\phi(r, z)}{r} = -\frac{dh}{d\Psi}. \qquad (6.42)$$

This equation is also obtained from (6.41) in the limit as $\Omega \to \infty$ with $(h(\Psi) - h_0)$ held constant.

Provided the circulation has everywhere a finite slope, we can represent $h(\Psi)$ as a polynomial in (Ψ/Ψ_e) where Ψ_e is the stream function at the slipstream edge:

$$h - h_0 = \sum_{m=0}^{M} a_m \left(\frac{\Psi}{\Psi_e}\right)^m. \qquad (6.43)$$

For most of the examples we found that a reasonable representation of a generic rotor circulation distribution is given by the simple polynomial

$$h - h_0 = a\left[\left(\frac{\Psi}{\Psi_e}\right) - \left(\frac{\Psi}{\Psi_e}\right)^2\right] \qquad (6.44)$$

and this has been found to give good agreement between the actuator disk theory and experiment. To evaluate (6.38)–(6.40), the slipstream vorticity is represented as a function of the form:

$$\frac{\omega_\phi(r, z)}{r} = \sum_{n=0}^{N} A_n(z) \left[1 - \left(\frac{r}{R(z)} \right)^2 \right]^n. \tag{6.45}$$

This allows the radial integrals to be performed using the integral below which plays a role in the actuator disk theory, analogous to the Glauert integral in elementary aerodynamics.

$$\int_0^R r'^{m+1} (R^2 - r'^2)^n J_m(sr')dr' = 2^n n! R^{m+n+1} s^{-(m+1)} J_{m+n+1}(sR). \tag{6.46}$$

The remaining integrals with respect to s are all of the form

$$I_{(\lambda,\mu,\nu)}(R(z'), r, z - z') = \int_0^\infty s^\lambda J_\mu(sR(z')) J_\nu(sr) e^{-|z-z'|} ds, \tag{6.47}$$

and for λ, μ, and ν integers, integrals of this form can always be evaluated in closed form in terms of complete elliptic integrals. This reduces Eqs. (6.38)–(6.40) to integrals representing the influence of an axial distribution of vortex disks, as illustrated in Fig. 6.11. After including the additional stream function due to the free-stream U_∞, Eq. (6.38) yields

$$\Psi(r, z) = \frac{U_\infty r^2}{2} + r \int_0^\infty \sum_{n=0}^{N} A_n(z') 2^{n-1} R^{2-n}(z') I_{(-(n+1),n+2,1)}(R(z'), r, z - z') dz'. \tag{6.48}$$

The coefficients $A_n(z)$ are weak functions of z and $R(z)$ and the $A_n(z)$ can be determined iteratively by employing Eqs. (6.41) and (6.43) and enforcing the boundary condition $\Psi(R(z), z) = \Psi_e$. From (6.48) this gives

$$\Psi_e = \frac{U_\infty R(z)^2}{2} + R(z) \int_0^\infty \sum_{n=0}^{N} A_n(z') 2^{n-1} R^{2-n}(z') I_{(-(n+1),n+2,1)}(R(z'), R(z), z - z') dz'. \tag{6.49}$$

The corresponding formulae for the axial and radial velocities are readily obtained. The azimuthal velocity component for a single-rotating propeller is given directly by

$$\frac{V_\phi(r, z)}{U_\infty} = \frac{h(\Psi(r, z)) - h_0}{\Omega r U_\infty}. \tag{6.50}$$

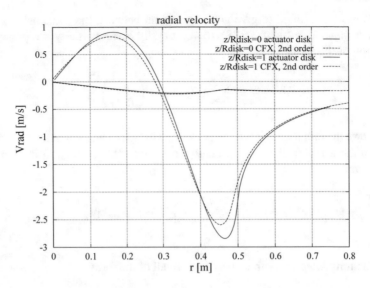

Fig. 6.12 Comparison of Conway's actuator disk theory to numerical solution of Euler's equation with load function of type (Eq. (6.44)) and $c_T = 0.95$ (Propeller case)

Reference [30] and from a different viewpoint [26] the results were compared with CFD-computations. The general expression was found to be very accurate except for the radial velocities close to the disk, which, in the numerical approach, has to be modeled with finite thickness (usually less then 10^{-2} times the disk radius, see Fig. 6.12).

6.6.2   Application to Wind Turbines

In [30, 31] the abovementioned ideas were applied to wind turbines, which in Conway's language corresponds to the negative sign case. Note that *advance ratio* J and tip speed ratio λ are related via

$$J = \frac{\pi}{\lambda} \, . \tag{6.51}$$

Streamlines for the full slip-stream iterated case with parabolic load $l(x) = -b \cdot (1 - r^2)$ with $b = -2.5$, corresponding to $c_T = 0.442$ and $c_P = 0.363$ are shown in Fig. 6.13. The slipstream expansion is from 0.97 (far upstream) to 1.11 (far downstream).

In addition Conway [7] was able to compare his exact method (when $J = 0$ corresponding to $\lambda \to \infty$) to a simple approximation using momentum theory for given shapes of wake deficits. Figure 6.14 shows examples of such far-wake velocity deficit distributions which are possible to model (except in the case of constant load).

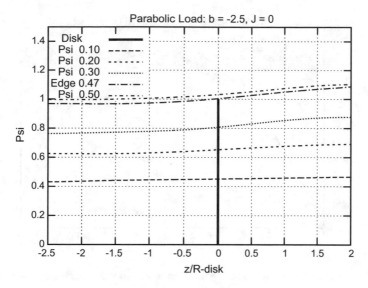

Fig. 6.13 Streamlines from Conway's actuator disk with parabolic load, b = 2.5

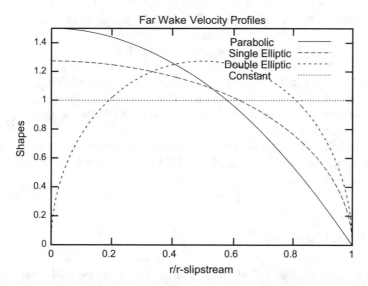

Fig. 6.14 Various shapes of far wake deficits

It is clearly seen that a constant velocity deficit has much more influence outside the hub.

Conway assumes

$$\left(\frac{R_{far} - Slipstream}{R_{tip}}\right)^2 = \frac{1 - a}{1 - 2a} \tag{6.52}$$

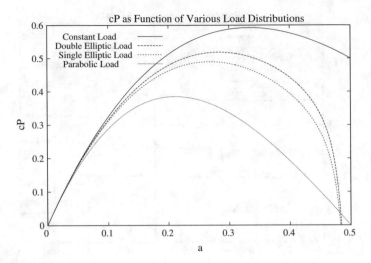

Fig. 6.15 Variation of c_P against induction factor from various wake deficit shape types from [7]

which becomes singular for $a \rightarrow 1/2$. For a parabolic-shaped deficit, it follows

$$c_{P,parabolic} = 4a \cdot (1 - a) \cdot (1 - 2a) . \tag{6.53}$$

The elliptic-shaped deficits are calculated in the same manner and all results are shown in Fig. 6.15. Close to $a = 0.5$ the variation of $c_P \rightarrow 0$ seem to indicate the breakdown of the simplified method. This is also confirmed when trying to perform a full slipstream iteration.

To summarize it may be said that full vortical slipstream seems to be in agreement with the numerical solution of the basic fluid equations, but so far no additional insights have be gained.

6.7   Optimum Rotors II

As we have seen, constant and variable circulation distributions lead to very different vortical flow patterns in the wake. To be more precise, we have for the Joukovski rotor when $\Gamma = const$ a single approximately 1D structure whereas in the Betz' case $\Gamma(r) = const. \cdot r^2/(\lambda^2 + r^2)$ a 2D vortex sheet. In Table 6.1 results from [21, 22] are compared from both cases together with Glauert's approach. As is clearly seen, the improvement is on the order of approximately 5%, but the practical feasibility of such a rotor remains undetermined. For any practical design, however, this is by far the simplest possible principle and was also deduced from momentum theory, presented in Sect. 5.3, by Sharpe (Chap. 5, [32]).

Table 6.1 Numerical values of c_P for optimal rotors, $B = 3$

TSR	Glauert	Joukovski	Betz/Goldstein	Improvement (%)
5	0.570	0.565	0.495	14
6	0.576	0.569	0.512	11
7	0.579	0.574	0.524	9
10	0.585	0.577	0.548	5
15	0.588	0.566	0.562	1

In the meantime further investigations [23, 28] indicate that slipstream expansion does not alter the results of Table 6.1 notably.

Case $\lambda = 7$ has been investigated in some detail by von Eitzen [9] with CFD using the Actuator Line (see Sect. 7.7) method implemented in the SOWFA distribution [3] and constant distribution of circulation along the blade was used. His results were $c_P = 0.578$ for the pure, and $c_P = 0.577$ for the implementation of Burton (Sect. 5.7, [8]) ($a = frac13, a' = \frac{2}{9}\lambda_r^{-1}$), which is remarkably close to Glauert's and Joukovskie's values.

Given $a = 1/3, a' = 0$ and restrict TSR > 3 a simple engineering equation for chord follows which sometimes is called *Betz' optimum blade shape*:

$$\frac{c}{R} \cdot \lambda^2 \cdot B \cdot c_L^{des} = \frac{16\pi}{9} \cdot x^{-1} \, , \quad x = r/R \, . \tag{6.54}$$

Interesting complimentary information can be found in Sect. 8.3 of [8] from Sect. 1.2.

6.8 Problems

Problem 6.1 Wilson's et al. [45] model, see Chap. 5.

A vertical axis wind turbine of *H-Darrieus* type may be analyzed with a model consisting of a simple rotating airfoil of length/height ($= H = 1$) and chord c. As the acting force we assume only lift with $c_L = 2\pi \cdot sin(\alpha)$. With help of Fig. 6.16 and ϕ being the circumferential angle show the following:

(a) The circumferential averaged axial induction is

$$a = \frac{B \, c}{2 R_{tip}} \frac{\Omega R_{tip}}{u_1} \, |cos(\theta)| \, . \tag{6.55}$$

(b) The averaged power of a rotor of B blades:

$$c_P = \pi x \left(\tfrac{1}{2} - \tfrac{4}{3\pi} x + \tfrac{3}{32} x^2 \right) \text{ with} \tag{6.56}$$

$$x \qquad := \sigma \lambda \qquad \text{and} \tag{6.57}$$

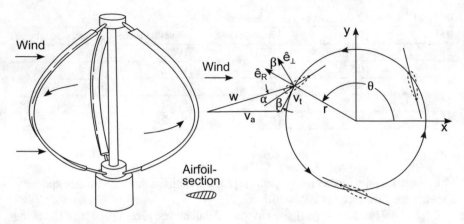

Fig. 6.16 H-Darrieus rotor

$$\sigma = \frac{Bc}{R_{tip}} \quad \text{and} \quad (6.58)$$

$$\lambda = \frac{\Omega R_{tip}}{u_1} \ . \quad (6.59)$$

(c) Show further that

$$c_{P,max} = 0.554 \text{ at} \quad (6.60)$$

$$\frac{x_{max}}{2} = a_{max} = 0.401 \ . \quad (6.61)$$

(d) Find a corresponding expression for the thrust-coefficient c_T.
(e) Discuss your findings!

Problem 6.2 Holmes Darrieus Model From [13]: Use thin-airfoil theory (Chap. 5 of Chap. 3, [34]) to show that the model of Problem 6.1 in the limit $B \to \infty$ together with $c \to 0$ may be transformed in a stationary flow problem using a vortex sheet of strength:

$$\gamma(\phi) = \frac{1}{2}ksq_R(\phi) \ . \quad (6.62)$$

Expanding

$$q_R(\phi) = \sum_{n=1}^{\infty} (a_n cos(n\phi) + b_n sin(n\phi)) \ . \quad (6.63)$$

For example, c_P may be shown to be

$$c_P = \sum_{n=1}^{\infty} (a_n^2 + b_n^2) \ . \quad (6.64)$$

Problem 6.3 Calculate the vorticity of the Scully vortex, Eq. (6.4).

Problem 6.4 Show that the flow field of Hill's spherical vortex, Eq. (6.5), obeys Euler's equation (3.31), provided that the pressure is given by

$$p = p_\infty - \frac{2A^2 a^7}{225} \left(\frac{r^2 - 2z^2}{\left(r^2 + z^2\right)^{3/2}} + \frac{a^3 \left(r^2 + 4z^2\right)}{\left(r^2 + z^2\right)^4} \right) . \tag{6.65}$$

Derive H inside the sphere $r^2 + z^2 < a^2$ by integrating

$$\nabla H = \mathbf{u} \times \omega \tag{6.66}$$

to

$$H = \frac{1}{10} A^2 r^2 (z^2 + r^2) . \tag{6.67}$$

Problem 6.5 Run the vortex patch code attached in Appendix A.6 to get comparable results as in Fig. 6.3.

References

1. Abramowitz M, Stegun I (1964) Handbook of mathematical functions. Dover Publications, Mineola
2. Chorin A (1994) Vorticity and turbulence. Springer, New York
3. Churchfield M, Lee S, Motuarty P (2012) Overview of the simulator for wind farm application (SOWFA), NREL
4. Conway JT (1995) Analytical solutions for the actuator disk with variable radial distribution of load. J Fluid Mech 297:327–355
5. Conway JT (1998) Exact actuator disk solution for non-uniform heavy loading and slipstream contraction. J Fluid Mech 365:235–267
6. Conway JT (1998) Prediction of the performance of heavily loaded propellers with slipstream contraction. CASJ 44:169–174
7. Conway JT (2002) Application of an exact nonlinear actuator disk theory to wind turbines. In: Proceedings of the ICNPAA, Melbourne, Florida, USA
8. Corten G (2001) Aspect ratio correction for wind turbine blades. In: Proceedings of the IEA joint action, aerodynamics of wind turbines, Athens, Greece
9. von Eitzen A (2018) Optimum circulation distribution on wind turbine rotor using SOWFA code. MSc thesis, The Danish Technical University, Lyngby, Denmark
10. Fukumoto Y, Okulov V (2005) The velocity field induced by a helical vortex tube. Phys Fluids 17:107101–107101-19
11. Goldstein S (1929) On the vortex theory of screw propellers. Proc R Soc Ser A 123:440–465
12. Helmholtz H (1858) Ueber Integrale der hydrodynamischen Gleichungen, welche den Wirbelbewegungen entsprechen (On integrals of the hydrodynamic equations which correspond to vortex motions). Crelle-Borchardt J f d reine und ang Mathematik LV:25–55 (Berlin, Preussen) (in German)
13. Holme O (1976) A contribution to the aerodynamic theory of the vertical-axis wind turbine. In: Proceedings of the international symposium on wind energy systems, Cambridge, UK, pp C4-55–C4-72

14. Joukovski N (1929) Théorie tourbillonnaire de L'hélice. Gauthier-Villars, Paris
15. van Kuik G (2013) On the generation of vorticity in rotor and disc flows. In: Proceedings of the ICOWES2013, Copenhagen, Denmark
16. Leweke et al T (2013) Local and global pairing in helical vortex systems. In: Proceedings of the ICOWES2013, Copenhagen, Denmark
17. Oeye S (1990) A simple vortex model. In: Proceedings of the 3rd IEAwind Annex XI, meeting, Harwell, UK
18. Okulov VL (2004) On the stability of multiple helical vortices. J Fluid Mech 521:319–342
19. Okulov VL (2010) Private communication
20. Okulov VL, Sørensen JN (2007) Stability of helical tip vortices in a rotor far wake. J Fluid Mech 576:1–25
21. Okulov VL, Sørensen JN (2008) Refined Betz limit for rotors with a finite number of blades. Wind Energy 11:415–426
22. Okulov VL, Sørensen JN (2010) Maximum efficiency of wind turbine rotors using Joukowsky and Betz approaches. J Fluid Mech 649:497–508
23. Okulov VL, Sørensen JN, van Kuik GAM (2013) Development of the optimum rotor theories. R&C Dynamics, Moscow-Izhevsk
24. Prandtl L (1918) Tragflügeltheorie. I. Mitteilung. Nachr d Königl Gesell Wiss zu Göttingen Math-phys Kl 451-477
25. Prandtl L (1919) Tragflügeltheorie. II. Mitteilung. Nachr d Königl Gesell Wiss zu Göttingen Math-phys Kl 107–137
26. Réthoré P-E, Sørensen NN, Zahle F (2010) Validation of an actuator disc model. In: Proceedings of the EWEC 2010, Brussels, Belgium
27. Saffman PG (1992) Vortex dynamics. Cambridge University Press, Cambridge
28. Segalini A, Alfredson PH (2013) A simplified vortex model of propeller and wind-turbine wakes. J Fluid Mech 725:91–116
29. Schaffarczyk AP (2011) Komponentenentwicklung und Konstruktion WIKO Urban 5, Aerodynamische Auslegung des Diffusors, des Blattes und eines passiven Pitch-Systems (Design and development of WIKO Urban 5: aerodynamic design of the diffusors, the blade and a passive pitch system), internal confidential report No. 81, Kiel, Germany (in German)
30. Schaffarczyk AP, Conway JT (2000) Comparison of nonlinear actuator disk theory with numerical integration including viscous effects. CASJ 46:209–215
31. Schaffarczyk AP, Conway JT (2000) Application of a nonlinear actuator disk theory to wind turbines. In: Proceedings of the 14th symposium on IEA joint action aerodynamics of wind turbines, Boulder, Co, USA
32. Strickland JH, Webster BT, Nguyen T (1979) A vortex model of the Darrieus turbine: an analytical and experimental study, Paper 79-WA/FE-6. ASME, New York
33. Tao T (2016) Finite time blowup for Lagrangian modifications of the three-dimensional Euler equation. Ann PDE 2:9
34. Theodorsen T (1948) Theory of propellers. McGraw-Hill, New York
35. Tibery CL, Wrench JW (1964) Tables of the Goldstein factor. Applied mathematics laboratory report, vol 1534. D. Taylor Model Basin, Washington
36. Wald QR (2006) The aerodynamics of propellers. Prog Aerosp Sci 42:85–128
37. Wu TY (1964) Flow through a heavily loaded actuator disc. Schiffstechnik 9(47):134–138
38. Wu J-Z, Ma H-Y, Zhou M-D (2006) Vorticity and vortex dynamics. Springer, Berlin
39. Young LA (2003) Vortex core size in the rotor near-wake, NASA/TM-2003-212275. Ames Research Center, Moffett Field, CA, USA

Chapter 7
Application of Computational Fluid Mechanics

You can't calculate what you haven't understood. (Originally thought to be from P. W. Anderson.)

7.1 Introduction

As we have seen in the previous chapters, due the nonlinear behavior it is very difficult—if not impossible—to get simple analytical solutions of the basic fluid dynamic equations in a systematic way. Therefore it has become normal to use (massive) numerical methods for solving them. In an ideal situation this would mean that only the Eqs. (3.37) from Chap. 3 are used (of course adapted to a suitable form for computers) together with a geometrical description of the problem (a wind turbine wing, for example) and a surrounding control volume for setting the boundary conditions for the unknown fields (pressure and velocity).

Unfortunately this is, even today, far too ambitious for most of the interesting engineering cases in which the Reynolds number are so high, that an important part of the flow is turbulent. By now CFD = Computational Fluid Dynamics, meaning numerical solving of any fluid mechanical problem, is an industry of its own and we must refer the reader to the specific literature for most of the details [1, 8, 9, 54].

As is a common practice, CFD is a three step process:

1. construct a 3D mesh from the geometry given (pre-processing),
2. solve the basic equations on (or in) this *finite volumes* (solving),
3. visualize and/or check the results (post-processing).

Everybody should know from undergraduate mathematics how to use at least two numerical schemes in calculus:

(a) finding zeros from the *false position method* and

(b) numerical 1D-integration by *Newton's Method* and/or *trapezoidal rule* and *Simpson's rule*.

© Springer Nature Switzerland AG 2020
A. P. Schaffarczyk, *Introduction to Wind Turbine Aerodynamics*,
Green Energy and Technology, https://doi.org/10.1007/978-3-030-41028-5_7

Primary output from any fluid dynamic calculation are the forces on the structure, in most cases on the blade. Because pressure is related *nonlocally* to the velocity field by Eq. (3.60) the whole 3D flow field must be calculated. It is immediately clear that a very high performance computer must be available if CFD for any fluid mechanical engine should be meaningful. Fortunately, since the first idea of using digital computers for (turbulent) flow calculations by Johann von Neumann [55], computing possibilities have been grown almost constantly by *Moore's law*, see Fig. (7.1). The energy consumption is enormous, today reaching more than 18 MW for the computer and cooling power of the same order of magnitude.

First CFD (RANS = Navier–Stokes Equations + empirical differential equations for turbulent field quantities, see Sect. 7.5) calculation appeared in the academic world around the end of the 1990s [11, 34]. Table 7.1 gives a personal overview to computer codes used by the author since the late 1980s. In the following discussion, we will try to introduce the reader to this technique as it applies to wind turbine blades.

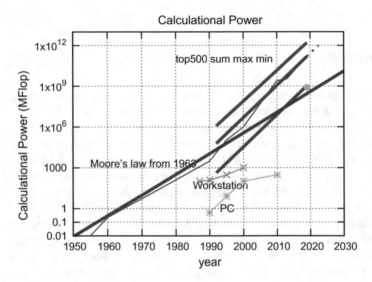

Fig. 7.1 Growth of computing power. Thin red line shows the development (absorbed by the middle blue line of TOP500's list) and the thick red one give Moore's extrapolation. Data from the TOP500 list is shown as three (sum, maximum, and minimum) blue lines. The dotted extension gives an indication when the fastest computer will reach exa10^{18}-flop-scale. The green dot gives an indication of computational power accessible for an ordinary user at a local HPC, occasionally just on top the minimum TOP500 list

Table 7.1 Some CFD codes

Code	Meshing	Special feature	Used by the author	Parallel computing	Language
Phoenics	1 Block structured	First code	1989–1993	No	FORTRAN77
CFX-4	Multiblock structured	Multiblocking	1995–2000	No	FORTRAN77
Fluent	Unstructured (U)	(U), easy to use	1997–	Yes	C
FLOWer	Multiblock	e^N method transition	2002–2012	Yes	FORTRAN77
Tau	Unstructured	(U)	2010–	Yes	C
OpenFOAM	Unstructured	Open source	2012–	Yes	C++

7.2 Pre-processing

Meshing a wing simply means putting a mesh around it as shown in Fig. 7.3. The
grid points usually define the locations where the field variables are calculated. The
distance of grid point should be chosen in that way that gradients are sufficiently
resolved to have a defined accuracy. But here is the problem: as we know (see Fig. 7.2)
sometimes pressure variation may be very large across small distances. To accurately
resolve these large gradients without wasting computational efforts—most portions
have very low gradients—one has to have an idea of the solution which the CFD
model will produce. This is clearly a contradiction. Even worse each reasonable
CFD computation has to be proven to be *grid independent*—that means that if the
grid is changed (finer or coarser) the solution must not change.

Mesh quality largely influences not only the accuracy but also the speed at which
this accuracy is reached.

Two major measures are used:

1. aspect ratio, meaning that all cell edges have similar lengths, see Fig. 7.8
2. orthogonality. See Fig. 7.8.

Figure 7.4 shows a mesh with mostly orthogonal cell which was coded by Trede
[53] along the lines of Sørensen [48]. Hyperbolic in this context means that the
generating 2D partial differential equation system is of hyperbolic type (like the
wave equation $u_{tt} - u_{xx} = 0$, the speed of the wave $c = 1$).

Three larger groups of mesh types are known, the last one being a mixture of the
first two:

1. Structured meshes meaning they have a fixed number of neighboring nodes, see
   Fig. 7.5,
2. Unstructured meshes with a free number of adjacent nodes, Fig. 7.7,
3. Mixtures of that, Fig. 7.6.

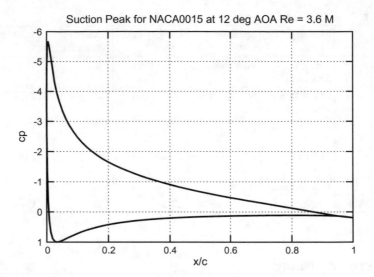

Fig. 7.2 Narrow suction peak close to the nose around a thin airfoil at large angle of attack

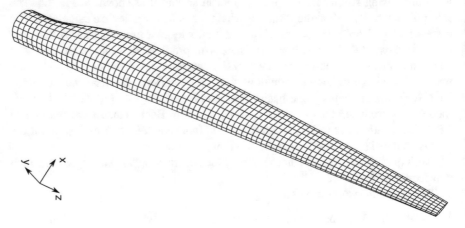

Fig. 7.3 Coarse mesh on a wind turbine wing

To summarize: mesh generation is still the most time-consuming part of a CFD job. Although unstructured grids are much easier and faster to generate than structured ones, often and in particular for boundary layer flow, structured grids still have to be used at least for specific parts.

A comparable new technique—the so-called *Chimera technique*—simply overlays two separate meshes on each other. Figure 7.29 shows an example of a wind turbine rotor consisting of three blades and a cylindrical hub.

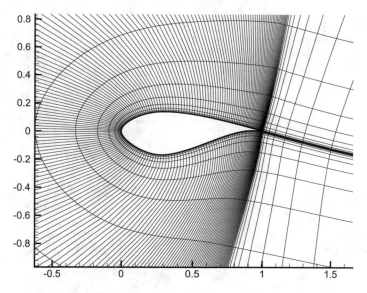

Fig. 7.4 Mesh perpendicular to a wind turbine wing, generated by a hyperbolic mesh generator [53]

Fig. 7.5 Mesh type 1:

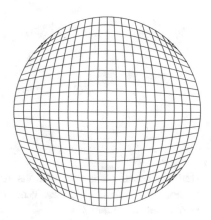

7.3   Solving the Numerical Equations

7.3.1   Discretization of Differential Equations

Discretization is to some extent the opposite procedure of the following *limits*:

$$\frac{dy(x)}{dx} = \lim_{\Delta x \to 0} \frac{\Delta y}{\Delta x} \ . \tag{7.1}$$

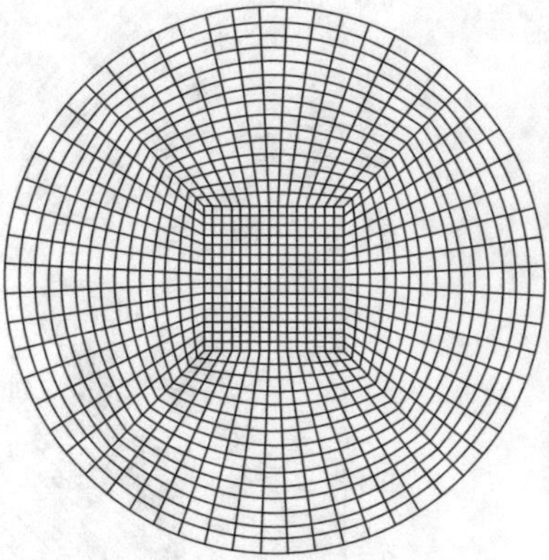

Fig. 7.6 Mesh type 2:
structured H-type mesh

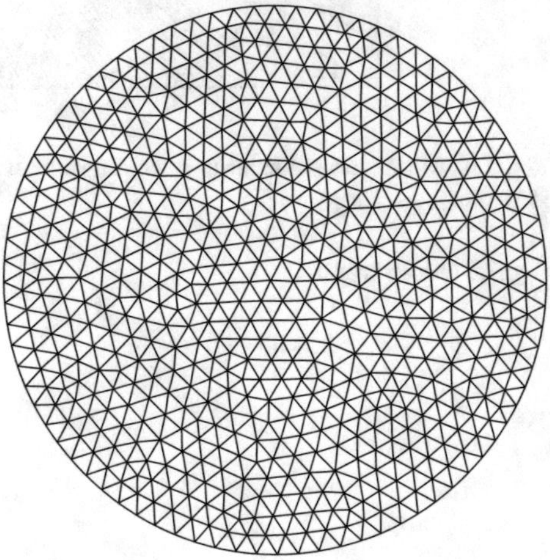

Fig. 7.7 Mesh type 3:
unstructured mesh

Methods based on generalizations for this trivial example, therefore, are called
method of *Finite Differences*. What on a glance seems to be easy to 1D cases is
it not at all in 2D and 3D.

Therefore, more sophisticated concepts based on the general conservation laws
discussed in Chap. 3 have been used to derive formulations suitable for digital computing.

7.3.2   *Boundary Conditions*

If the computational domain could be infinitely large, the only *boundary conditions*
(BoCos) would be $\mathbf{u} = const$ and $p_0 = 0$. At solid walls it would be simply $\mathbf{u} = 0$.
Other BoCos were introduced to decrease the extent of the computational domain,
for example bi-symmetry considerations.

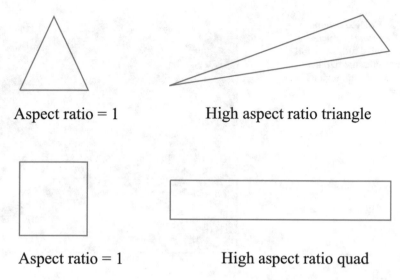

Fig. 7.8 Orthogonality and aspect ratio for triangles (upper row) and quads (lower row)

As the flow must be forced at least from one part of the domain in many cases *inflow BoCos* are used. They are very similar to the abovementioned *far field* BoCos.

Complementary to inflow conditions the conditions defining an *outlet* simply specify that the flow does not change anymore, meaning that all gradients are set to zero.

The influence of BoCos both on the numerical implementation and on the accuracy of the result may not be underestimated. For example, a wind turbine blade (modeled as a 120° segment with *periodic boundary conditions*) uses a far field domain of 10–20 rotor diameters.

7.3.3   Numerical Treatment of the Algebraic Equations

After everything has been reduced to a huge number of discrete equations, efficient numerical algorithms are needed. The impressive improvement in speed and size of CFD calculations originated partly from increased computer power (as shown in Fig. 7.1) and partly in more efficient solving algorithms. As a well-known example, the multigrid algorithm is mentioned here.

Fig. 7.9 Colored contour plot of pressure coefficient around a wind turbine airfoil red: high pressure regions, blue: low pressure regions, AOA = 5°

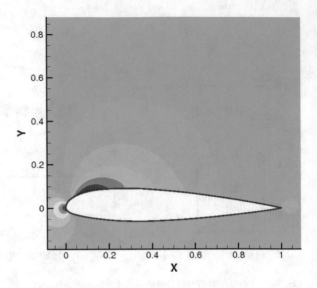

7.4   Post-Processing: Displaying and Checking the Results

Once a *converged solution* which is a solution with a prescribed tiny change (typically 10^{-6} or at least 10^{-3}) after one last iteration is reached, the problem of checking the results starts. It used to be done by colored plots for velocity and pressure fields (see Fig. 7.9) for stationary flow, but with the emergency of unsteady methods like LES (Sect. 7.5.12) movies have become more fashionable. Any achievement of accuracy seems to be a delicate and highly nontrivial task. We will come back to this in more detail in Sect. 7.8.2.

7.5   Turbulence Models for CFD

Turbulence (and now more and more transition) modeling is at the core of CFD. If the flow is laminar, CFD is capable of generating as accurate numerical solutions as desired. This is (unfortunately) not at all the case for turbulent flow, which in almost all cases interesting for wind turbine applications is present. Therefore, it has to be modeled in an efficient way suitable for numerical solving of nonlinear partial differential equations of the *transport equation* type.

Most models start from the statistical approach of Reynolds averaging and therefore CFD is in most cases *RANS* = Solution of Reynolds Averaged Navier–Stokes Equations. Splitting into average and deviation using

$$u_\alpha := U_\alpha + u'_\alpha \, , \tag{7.2}$$

$$< u'_\alpha > \, = 0 \tag{7.3}$$

we have further

$$\partial_t U_\alpha + U_\beta \partial_\beta U_\alpha = -\partial_\alpha p + \partial_\beta \left(\nu S_{\alpha\beta} - < u'_\alpha u'_\beta > \right) . \tag{7.4}$$

In Eq. (7.4) the so-called *Reynolds stresses*

$$\tau_{\alpha\beta} := < u'_\alpha u'_\beta > \tag{7.5}$$

are introduced which display the influence of velocity correlations. These quantities have to be precisely modeled. In addition, $S_{\alpha\beta}$ from Eq. (3.35) is defined as the *deformation tensor*, and is comprised of averaged velocities.

7.5.1 Prandtl's Mixing Length Model

One of the first empirical models of turbulence came from Prandtl [29, 56] and was inspired by kinetic gas theory. It was postulated that *lumps of turbulence* (*Turbulenzballen* in German) are mixed up after a certain length ℓ:

$$< u'_\alpha u'_\beta >:= -\ell^2 \left(\partial_\beta U_\alpha \right) \tag{7.6}$$

Modeling now has to be applied to ℓ. Karman proposed close to a wall: $\ell = \kappa \cdot y^+$ with $\kappa \approx 0, 4$. Quite successful was the application to jets [56].

An ongoing discussion is about the so-called *universality* of κ meaning that it is a constant of turbulent flow independent of special boundary conditions.

Generalizations of the mixing length model are as follows:

- the already mentioned von Karmans approach inside boundary layers $\ell = \kappa y^+$,
- *van Driest's* boundary layer damping

$$\ell_{mix} = \kappa y(1 - \exp(-y^+/A_0^+)) \tag{7.7}$$
$$\text{with} \qquad A_0^+ \approx 26, \tag{7.8}$$

- Cebessi–Smith's two-layer model,
- Smagorinski's *sub-grid scale model* for LES (see Sect. 7.5.12),
- the Baldwin–Lomax model with $\nu_T = \ell_m^2 \cdot \sqrt{\omega_\beta \omega_\beta}$ with $\boldsymbol{\omega} = \nabla \times \mathbf{u}$, the vorticity vector.

7.5.2 One Equation Model

Equation 7.6 is called an algebraic turbulence model, because only one turbulent quantity is connected to the averaged flow by an algebraic equation. Applied to jets it gave some accurate results, but it fails for more complex situations. Therefore, it seems natural to proceed with the diagonal component of the stress tensor, the turbulent kinetic energy from Eq. (3.120).

The equation for the turbulent kinetic energy derived from the Navier–Stokes equation reads as follows:

$$\partial_t k + U_\beta \partial_\beta k = \qquad (7.9)$$

$$\tau_{\alpha\beta} \partial_\beta U_\alpha - \varepsilon + \partial_\beta \left(\nu \partial_\beta k - \frac{1}{2} \underbrace{< u'_\alpha{}^2 u'_\beta >}_{*} - \underbrace{< p' u'_\beta >}_{**} \right), \qquad (7.10)$$

here ε from Eq. (7.11) defines turbulent dissipation.

$$\varepsilon := \frac{\nu}{2} \cdot \sum_\alpha \sum_\beta \left\langle \frac{\partial u'_\alpha}{\partial x_\beta} + \frac{\partial u'_\beta}{\partial x_\alpha} \right\rangle^2. \qquad (7.11)$$

Equation (7.11) is not closed, because of triple-correlations (*) $< u'_\alpha{}^2 u'_\beta >$ and velocity–pressure correlations.

Further we need a link between ε and k.

If we set

$$(*) + (**) := -\frac{\mu_T}{\sigma_k} \partial_\beta k \qquad (7.12)$$

$$\varepsilon := C_D k^{3/2} / \ell \qquad (7.13)$$

we arrive a single differential equation model:

$$\partial_t k + U_\beta \partial_\beta k = \qquad (7.14)$$

$$\tau_{\alpha\beta} \partial_\beta U_\alpha - \varepsilon + \partial_\beta \left(\nu + \mu_T / \rho \sigma_k \partial_\beta k \right) \qquad (7.15)$$

$$\text{with } \mu_t = \rho k^{1/2} \ell \qquad (7.16)$$

$$+ \text{ empirical model for } \ell. \qquad (7.17)$$

Equation (7.17) is the weak link. Again we have to introduce empirical approximation which in may cases are from dimensional analysis, together with new constants which have to be determined experimentally.

7.5.3 Spalart–Allmeras Model

This model is *popular* in 2D airfoil aerodynamics. It starts from a turbulent viscosity which is given by

$$v_t = \tilde{v} f_{v1}, \quad f_{v1} = \frac{\chi^3}{\chi^3 + C_{v1}^3}, \quad \chi := \frac{\tilde{v}}{v}. \tag{7.18}$$

The transport equations for the modified viscosity $\tilde{v}$ are

$$\frac{\partial \tilde{v}}{\partial t} + u_j \frac{\partial \tilde{v}}{\partial x_j} = \tag{7.19}$$

$$C_{b1}[1 - f_{t2}]\tilde{S}\tilde{v} + \frac{1}{\sigma}\{\nabla \cdot [(v + \tilde{v})\nabla\tilde{v}] + C_{b2}|\nabla v|^2\} - \tag{7.20}$$

$$\left[C_{w1} f_w - \frac{C_{b1}}{\kappa^2} f_{t2}\right]\left(\frac{\tilde{v}}{d}\right)^2 + f_{t1}\Delta U^2 \tag{7.21}$$

$$\tilde{S} \equiv S + \frac{\tilde{v}}{\kappa^2 d^2} f_{v2}, \quad f_{v2} = 1 - \frac{\chi}{1 + \chi f_{v1}} \tag{7.22}$$

$$f_w = g\left[\frac{1 + C_{w3}^6}{g^6 + C_{w3}^6}\right]^{1/6}, \quad g = r + C_{w2}(r^6 - r), \quad r \equiv tu\frac{\tilde{v}}{\tilde{S}\kappa^2 d^2} \tag{7.23}$$

$$f_{t1} = C_{t1} g_t \exp\left(-C_{t2}\frac{\omega_t^2}{\Delta U^2}[d^2 + g_t^2 d_t^2]\right) \tag{7.24}$$

$$f_{t2} = C_{t3} \exp\left(-C_{t4}\chi^2\right). \tag{7.25}$$

The rotation tensor $S = \sqrt{2\Omega_{ij}\Omega_{ij}}$ uses Ω_{ij} which is given by

$$\Omega_{ij} = \frac{1}{2}(\partial u_i/\partial x_j - \partial u_j/\partial x_i) \tag{7.26}$$

and d is the distance from the closest surface.

The model's constants are as follows:

$$\begin{array}{rcl}
\sigma & = & 2/3 \\
C_{b1} & = & 0.1355 \\
C_{b2} & = & 0.622 \\
\kappa & = & 0.41 \\
C_{w1} & = & C_{b1}/\kappa^2 + (1 + C_{b2})/\sigma \\
C_{w2} & = & 0.3 \\
C_{w3} & = & 2 \\
C_{v1} & = & 7.1 \\
C_{t1} & = & 1 \\
C_{t2} & = & 2 \\
C_{t3} & = & 1.1 \\
C_{t4} & = & 2
\end{array} \tag{7.27}$$

The reader may decide by himself or herself if a model with such a large number of constants seems to be general relevance.

7.5.4 Two Equation Models

Kolmogorov [15] and Prandtl [30] (see also the paper of Spalding [52]) independently introduced a set of differential equations. Both emphasized the importance of k, the turbulent kinetic energy, but only one introduced a second quantity to use, whereas Prandtl stayed at one equation, Kolmogorov uses a kind of a frequency ω as an additional field.

As the k-equation (Eq. 3.120) includes the turbulent dissipation ε one may hope to deduce a more realistic model if one use an additional differential equation for that quantity. ε was defined in Eq. (7.11). From NSE we may deduce [56]

$$(\partial_t + u_\beta \partial_\beta)\varepsilon = \quad (7.28)$$
$$-2\left(<\partial_\gamma u_\alpha \partial_\gamma u_\beta> + <\partial_\alpha u_\gamma \partial_\beta u_\gamma>\right)\partial_\beta u_\alpha - 2\nu <u_\gamma \partial_\beta u_\alpha> \partial_{\beta\gamma} u_\alpha \quad (7.29)$$
$$-2\nu <\partial_\gamma u_\alpha \partial_\delta u_\alpha \partial_\gamma u_\beta> -2\nu^2 <\partial_{\gamma\delta} u_\alpha \partial_{\gamma\delta} u_\alpha> \quad (7.30)$$
$$+\partial_\beta \left[\nu \partial_\beta \varepsilon - \nu <u_\beta \partial_\delta u_\alpha \partial_\delta u_\alpha> -2\nu/\rho <\partial_\delta p \partial_\delta u_\beta>\right] \quad (7.31)$$

Apparently this equation for ε is much more sophisticated than that for k. In more detail

1. Equation (7.29): production (of ε)
2. Equation (7.30): dissipation (of ε)
3. Equation (7.31): diffusion and turbulent transport (of ε).[1]

7.5.5 Standard k-ε-Model

This model [26] sometimes is described as the *workhorse* of CFD and is one of the earliest models applied to CFD. Usually it is defined by

$$\tau_{\alpha\beta} = \frac{2}{3}k\delta_{\alpha\beta} - \nu_T S_{\alpha\beta} \quad (7.32)$$

$$\nu_T = C_\mu \frac{k^2}{\varepsilon} \quad (7.33)$$

$$\partial_t k + U_\alpha \partial_\alpha k = -\tau_{\alpha\beta}\partial_\beta - \varepsilon + \partial_\alpha \left(\frac{\nu_T}{\sigma_K}\partial_\alpha k\right) + \nu \partial_{\alpha\alpha} k \quad (7.34)$$

$$\partial_t \varepsilon + U_\alpha \partial_\alpha \varepsilon = \underbrace{-C_{\varepsilon 1}\frac{\varepsilon}{k}\tau_{\alpha\beta}\partial_\beta U_\alpha}_{(1)} \underbrace{-C_{\varepsilon 2}\frac{\varepsilon^2}{k}}_{(2)} + \quad (7.35)$$

---

[1]Quadratic correlations may be replaced by diffusions term because of the *Fluctuation–Dissipation Theorem* from linear and equilibrium statistical mechanics ([54] in Chap. 3). It states that the variance of equilibrium fluctuations determines the strength of losses by small disturbances as well.

$$\underbrace{\partial_\alpha \left(\frac{\nu_T}{\sigma_\varepsilon}\partial_\alpha\varepsilon\right) + \nu\partial_{\alpha\alpha}\varepsilon}_{(3)}$$

and closure constants:

$$C_\mu = 0.09 \,,\, C_{\varepsilon 1} = 1.44 \,,\, \text{and } C_{\varepsilon 2} = 1.92; \sigma_K = 1.0; \sigma_\varepsilon = 1.3 \qquad (7.36)$$

Terms (1), (2), and (3) in the ε-equation, Eq. (7.36) are the most unsavory part of the k-ε-model.

As a guiding line to truncate the higher order correlations of $u_\alpha, p, k, \varepsilon$ only dimensional correctness is used and newly emerging *closure constants* (7.36) which have to be determined by experiment. Which one have to use is discussed in [56]. As this model has been used in the early time of CFD, in non-aerospace applications a number of other different once emerged, one of them being the k-ω-model and derivative.

7.5.6 *k-ε-Model from Renormalization Group (RNG) Theory*

A method well know from quantum field theory and later applied to quantum and classical non-relativistic many body systems was also applied to turbulence, see [45] in Chap. 3. While there is still some discussion about the range of validity to turbulence, its outcome was a modified k-ε-model [42, 59]:

$$\partial_t k + u_\alpha \partial_\alpha k = \tau_{\alpha\beta}\partial_\beta U_\alpha - \varepsilon$$
$$+ \partial_\alpha(\alpha(\nu_T)\nu\partial_\alpha k) \qquad (7.37)$$
$$\partial_t \varepsilon + u_\alpha \partial_\alpha \varepsilon = \frac{\nu_T}{2}\sqrt{\varepsilon}C_c Y_\varepsilon \tau_{\alpha\beta}\partial_\beta U_\alpha$$
$$+ \varepsilon^{3/2}Y_\varepsilon + \partial_\alpha(\alpha(\nu_T)\nu\partial_\alpha\varepsilon). \qquad (7.38)$$

In addition, the turbulent viscosity ν_T is being set into connection with quantities like $k/\sqrt{\varepsilon}$, Y_ε, and σ_K.

$$\frac{d}{d\nu_T}\frac{k}{\sqrt{\varepsilon}}(\nu_T) = 1.72 \, (\nu_T^3 + C_c - 1)^{-1/2 \cdot \nu_T} \qquad (7.39)$$

$$\frac{dY_\varepsilon}{d\nu_T} = -0.5764 \, (\nu_T^3 + C_c - 1)^{-1/2} \qquad (7.40)$$

$$\frac{1}{\nu_T} = \left(\frac{1.3929 - \alpha(\nu_T)}{0.3929}\right)^{0.63}\left(\frac{2.3929 - \alpha(\nu_T)}{3.3929}\right)^{0.37} \qquad (7.41)$$

and constants:

$$C_\mu = 0.084; C_{\varepsilon 1} = 1.063; C_{\varepsilon 2} = 1.72;$$
$$\sigma_K = 0.7179; \sigma_\varepsilon = 0.7179; C_c \approx 75. \qquad (7.42)$$

Further discussion [42, 43] clarified some of the very controversially discussed items, especially on the status of the ε-equation (7.38) and the model (closure) constants.

7.5.7 Boundary Conditions for Turbulent Quantities Close to Walls

Any solution of a partial differential equation has to be supplemented by *boundary conditions*. Special attention must be spent on the new quantities like k and ε. Applicable boundary conditions are as follows:

$$\lim_{y^+ \to 0} k(y^+) = \qquad (7.43)$$

$$= \lim_{y^+ \to 0} \varepsilon(y^+) = 0 \,. \qquad (7.44)$$

From DNS [49] one may compare the standard model and others. See Figs. 7.10 and 7.11.

From that we see, that close to the wall k^+ has a maximum at approximately $y^+ \approx 20$ but ε^+ seems to reach a constant value only very close to the wall.

In addition, one interpolates velocity field by the well-known logarithmic velocity profile, Eq. (3.149). It has to be noted that meshing depends on some of the turbulence models used.

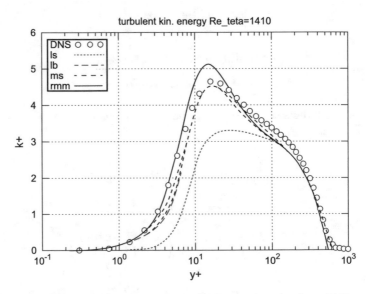

Fig. 7.10 Turbulent kinetic energy close to a wall for a flat plate boundary layer. DNS means direct numerical simulation, ls means the standard k-ε model. Adapted from [25]

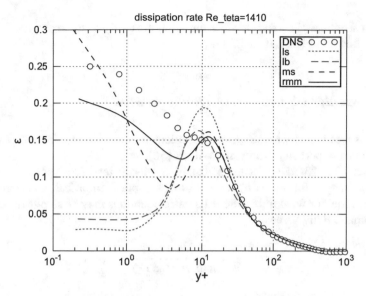

Fig. 7.11 Turbulent energy dissipation close to a wall. Same description as in Fig. 7.10

To summarize we may list some of the drawbacks of the original k-ε-model:

1. only fully turbulent flow may be investigated,
2. no consistent extrapolation down to $y^+ \rightarrow 0$ possible,
3. so-called interpolating wall functions are valid only for non-separated flow.

   Another drawback is that the k-ε-model may become inaccurate if strong inertia forces become important when used in rotating frames of reference.

7.5.8 Low-Re-Models

Some of the abovementioned turbulence expressions do not produce accurate results close to walls. Therefore, correction for this so-called *Low-Reynolds number* have been developed [56]. A useful definition for a turbulent Re number may be given in the following equation:

$$Re_t = \frac{\sqrt{k}\ell}{\nu} = \frac{k^2}{\nu_\varepsilon} \cdot \frac{\nu_t}{\nu} . \tag{7.45}$$

$Re_t < 100$ is **not** attained:

1. shortly after transition,
2. in heavily accelerated flow,
3. in the far field of wakes,
4. small wall distance.

In most cases, damping- and blending-functions are used and some progress has been seen as shown in Figs. 7.10 and 7.11.

7.5.9 Menter's Shear Stress Transport Model

The aerospace industry uses ω (a local turbulent angular velocity) instead of ε as the second most important quantity, see Wilcox [56].

Menter [22, 23] developed a so-called S(hear)S(tress)T(ransport)-k-ω model which proved useful and is by now the most popular empirical engineering turbulence model for wind turbine flow. Its basic equations may be summarized as

Kinematic Eddy Viscosity:

$$\nu_T = \frac{a_1 k}{\max(a_1 \omega, S F_2)} , \tag{7.46}$$

Turbulence Kinetic Energy:

$$\frac{\partial k}{\partial t} + U_j \frac{\partial k}{\partial x_j} = P_k - \beta^* k\omega + \frac{\partial}{\partial x_j}\left[(\nu + \sigma_k \nu_T)\frac{\partial k}{\partial x_j}\right], \tag{7.47}$$

Specific Dissipation Rate:

$$\frac{\partial \omega}{\partial t} + U_j \frac{\partial \omega}{\partial x_j} = \alpha S^2 - \beta\omega^2 + \frac{\partial}{\partial x_j}\left[(\nu + \sigma_\omega \nu_T)\frac{\partial \omega}{\partial x_j}\right] + 2(1 - F_1)\sigma_{\omega 2}\frac{1}{\omega}\frac{\partial k}{\partial x_i}\frac{\partial \omega}{\partial x_i}, \tag{7.48}$$

Closure Coefficients and Auxiliary Relations:

$$F_2 = \tanh\left[\left[\max\left(\frac{2\sqrt{k}}{\beta^* \omega y}, \frac{500\nu}{y^2 \omega}\right)\right]^2\right] \tag{7.49}$$

$$P_k = \min\left(\tau_{ij}\frac{\partial U_i}{\partial x_j}, 10\beta^* k\omega\right) \tag{7.50}$$

$$F_1 = \tanh\left\{\left\{\min\left[\max\left(\frac{\sqrt{k}}{\beta^* \omega y}, \frac{500\nu}{y^2 \omega}\right), \frac{4\sigma_{\omega 2} k}{C D_{k\omega} y^2}\right]\right\}^4\right\} \tag{7.51}$$

$$C D_{k\omega} = \max\left(2\rho\sigma_{\omega 2}\frac{1}{\omega}\frac{\partial k}{\partial x_i}\frac{\partial \omega}{\partial x_i}, 10^{-10}\right) \tag{7.52}$$

$$\phi = \phi_1 F_1 + \phi_2(1 - F_1) \tag{7.53}$$

$$\alpha_1 = \frac{5}{9}, \alpha_2 = 0.44 \tag{7.54}$$

$$\beta_1 = \frac{3}{40}, \beta_2 = 0.0828 \tag{7.55}$$

$$\beta^* = \frac{9}{100} \tag{7.56}$$

$$\sigma_{k1} = 0.85, \sigma_{k2} = 1 \tag{7.57}$$

$$\sigma_{\omega 1} = 0.5, \sigma_{\omega 2} = 0.856. \tag{7.58}$$

7.5.10 Reynolds Stress Models

From NSE we may even derive a complete set of equations for all of the Reynolds stresses Eq. (7.5):

$$\partial_t \tau_{\alpha\beta} + U_\gamma \partial_\gamma \tau_{\alpha\beta} =$$

$$-\tau_{\alpha\gamma} \partial_\gamma U_\beta - \tau_{\beta\gamma} \partial_\gamma U_\alpha \tag{7.59}$$

$$+\Pi_{\alpha\beta} - \varepsilon_{\alpha\beta} - \partial_\gamma \dot{C}_{\alpha\beta\gamma} + \nu \partial_{\gamma\gamma} \tau_{\alpha\beta} \tag{7.60}$$

with :

$$-\tau_{\alpha\gamma} \partial_\gamma U_\beta - \tau_{\beta\gamma} \partial_\gamma U_\alpha$$

Stress creation

$$\Pi_{\alpha\beta} = \langle p \partial_\beta u_\alpha + \partial_\alpha u_\beta \rangle \tag{7.61}$$

pressure-dilation correlation

$$\varepsilon_{\alpha\beta} = 2\nu \langle \partial_\gamma u_\alpha \partial_\gamma u_\beta \rangle \tag{7.62}$$

dissipation rate correlation and

$$C_{\alpha\beta\gamma} = < u_\alpha u_\beta u_\gamma > + < p u_\alpha > \delta_{\beta\gamma} + < p u_\beta > \delta_{\alpha\gamma} \tag{7.63}$$

diffusions correlation of third order. $\tag{7.64}$

Compared to Eq. (3.120) these equations are far more complicated and use much more than 20—sometimes difficult to determine—new constants.

7.5.11 Direct Numerical Simulation

Direct numerical simulation (DNS) with no turbulence modeling at all is possible only for very few cases. Reference [12] gives an account of the work on isotropic turbulence ($Re_\lambda = 1131$) on a 4096^3 grid.

Data for a 1024^3 ($Re_\lambda = 433$) grid is available even for public [16].

For wall-bounded flow only the simulation of Spalart [49] for a turbulent boundary layer on a flat plate $Re_\theta = 1410$ corresponding to $Re_x \approx 10^6$ is available. Some of his findings are shown in Figs. 7.10 and 7.11 and are compared to simpler RANS models there.

The Exa FLOW project (Enabling Exascale Fluid Dynamics), coordinated by The Royal Institute of Technology—KTH, Sweden. These systems will have up to 10^9 processors and it is expected that Reynolds number for HIT (homogeneous isotropic turbulence) will increase by a factor of 3–5 to about 7000. Spalart [50] conjectured 20 years ago that a full turbulence model-free CFD simulation of an airplane would need 1016 grid points and $10^{7.7}$ time steps. As seen from today, it seems were likely to witness this event much earlier on an Exascale computer.

7.5.12   Large and Detached Eddy Simulation

An intermediate concept between DNS and RANS is *Large Eddy Simulation*, where a different concept of separation of velocities is used:

$$u = \bar{u} + u' , \tag{7.65}$$

$$u, \bar{r} t = \int G(r - \xi; \Delta) \cdot u(\xi, t) \, d\xi . \tag{7.66}$$

An important difference between averaging and filtering is

$$\bar{\bar{u}} \neq \bar{u} . \tag{7.67}$$

Δ is the usual filter size. As in the case of RANS the smallest scales (typically several ten times larger than the Kolmogorov scale) have to be modeled. Starting point is the so-called Smagroinsky *sub-grid scale stress, SGS*

$$\tau_{ij} = \mu_t \cdot \frac{1}{2} \left(\frac{\partial \bar{u}_i}{\partial x_j} + \frac{\partial \bar{u}_j}{\partial x_i} \right) \tag{7.68}$$

$$\mu_t = \rho C_s^2 \Delta^2 \sqrt{S_{ij} S ij} \quad \text{and} \quad C_S = 0.1 \dots 0.24 . \tag{7.69}$$

Some properties of LES are as follows:

- LES is always unsteady;
- LES is 3D;
- Near walls the eddy viscosity has to be damped out.

To compare the computational effort (usually against DNS) we may refer to Pope, Problem 13.29, [54] in Chap. 3.

The number of cells N^3 for DNS scales as $\approx 160 \cdot Re^3$ and for LES (depending of the *near wall resolution* $\sim Re^{1.8}$. Both calculations are transient. For DNS, the number of time steps needed is roughly $15 \cdot Re_\lambda$ and for LES one order of magnitude smaller.

The total number of cells is smaller only by a factor of 10–20 and the CPU time is one order of magnitude smaller.

If solid boundaries for any part of the application are avoided (by means of actuator disk and actuator line methods, see Sect. 7.7) a major application is wake-aerodynamics of wind turbines and even wind farms [32, 58].

An interesting mixture between LES and RANS, *Detached Eddy Simulation*, [51] may be used especially for highly separated flow.

7.6 Transitional Flow

CFD Codes in general only model fully turbulent flow. Laminar parts have to be separated *by hand*. As we have seen in Chap. 5 (Fig. 5.13) reasonable c_P-values around 0.5 need lift-to-drag ratios of more that 80. Keeping c_L fixed to moderate values around 1, the only chance is to decrease drag.

In the pure laminar case Blasius (see Eq. 3.102) has shown that

$$c_D = \frac{1.3}{\sqrt{Re_x}}. \tag{7.70}$$

whereas for the fully turbulent case

$$c_D = 2 \left[\frac{\kappa}{\ln Re_x} \right]^2. \tag{7.71}$$

Therefore, simulation of high lift-to-drag ratios of airfoils necessarily relies on modeling all laminar, transitional, and turbulent parts. In [17, 24, 47, 57] some of the details of currently used methods and approaches are described (Fig. 7.12).

It has to noted that at least two important so-called *scenarios* are used to describe the mechanisms from laminar to turbulent flow on a wall.

- Tollmien–Schlichting and
- Bypass [37].

The first one already described in Chap. 3 may be used for inflow conditions with less than 1% turbulence intensity. If more turbulence is present this transition schema is *bypassed*.

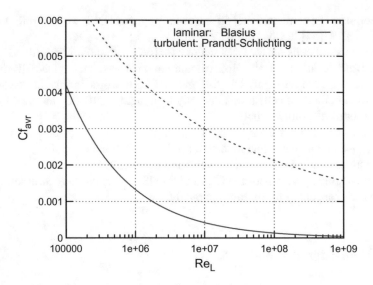

Fig. 7.12 Comparison of analytical approximation, see also Fig. 3.16

Much effort has been spent to describe and explain the findings during the NASA-Ames Phase VI experiments (Chap. 8) [39], see Sect. 8.5. Figures 7.21 and 7.20 show comparisons for the S809-airfoils used and Figs. 7.22 and 7.23 show the influence on the power curve.

During the IAEwind Task 29 (MexNext) Phase 3 (2015–2017) extensive investigations [19, 33] to detect laminar-turbulent from MEXICO (2006) and NEW MEXICO (2014) experiments were undertaken. Result are shown in Fig. 7.13. It can be seen that using a transition model at least for the 15 m/s case gives results closer to the measured values. Nevertheless, deviations as well as error bars are still in the 5% range (Fig. 7.14).

7.7   Actuator Disk and Actuator Line Modeling of Wind Turbine Rotors

7.7.1   Description of the Model and Examples

7.7.1.1   Actuator Disk Models

As a first example of the usefulness of CFD we present results using Navier–Stokes (or Euler) solvers when combined with Actuator Disk method [20, 41]. This method uses the full 3D (not 2D as in the simple Blade-Element-Momentum method from Chap. 5) flow field as input. It is particularly suited for the comparison of free-field

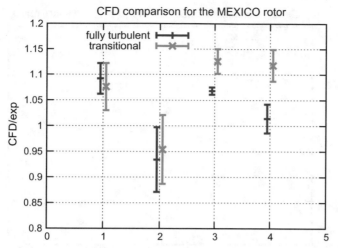

Fig. 7.13 Comparison of fully turbulent and transitional CFD simulation with experiment. From [33]

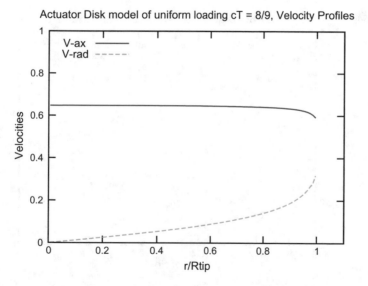

Fig. 7.14 Example of CFD-based actuator disk model, ax = axial, rad = radial

configurations and more complicated structures surrounding the turbine have to be modeled (Figs. 7.15 and 7.16).

As a test, we show in the next figures results for the Betz' rotor ($c_T = 8/9$) and Glauert's rotor ($\lambda = 2$).

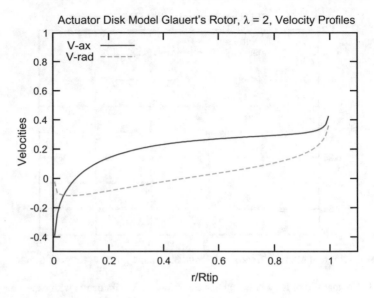

Fig. 7.15 Example of application of the actuator disk method to the Betz rotor

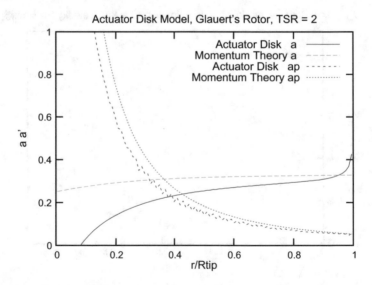

Fig. 7.16 Example of application of the actuator disk method to the Glauert rotor

It is clearly seen that radial velocity components are not negligible and have strong influence toward the edge of the disk.

Figure 7.17 shows the flow field of the Rotor for $v_{in=10}$ m/s with and without tunnel [20]. It is clearly seen that there is a strong interaction between the wind turbine and the wind tunnel.

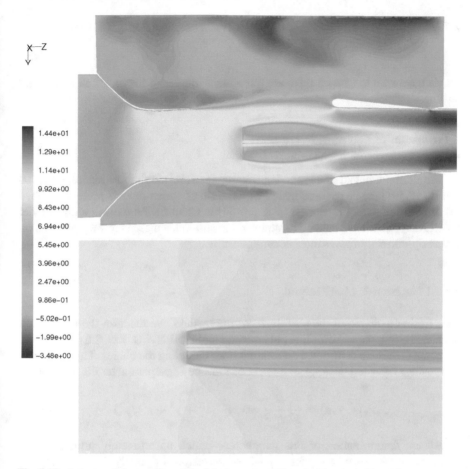

Fig. 7.17 Actuator disk modeling of the Mexico rotor with and without wind tunnel

Shen et al. [41] discuss the same measured data with their actuator line method, also with and without a wind tunnel. They found an influence of the order of 3% to the axial velocity in the rotor plane. Knowing that thrust $\sim v^2$ and power $\sim v^3$ we may see that variations in these quantities magnifie to about 6% or even 9%, respectively.

The MEXICO experiment is described in more detail in Chap. 5 [29] and Sect. 8.6. Comparison with numerical results provided important additional information on both experimental as well as theoretical accuracy.

We close this section with remark that the actuator disk/line method is particularly useful when combined with LES simulation for wake investigations of single turbines and/or wind farms.

7.7.2 Improved Models for Finite Numbers of Blades in BEM Codes

As we have seen in Chap. 5 there are three main sources of losses for an ideal wind turbine:

- Swirl: > 0,
- Finite number of blades: $B < \infty$,
- Drag of lift-generating airfoils: $d/L > 0$.

The all important ad-hoc Prandtl correction for finite numbers of blades, Fig. 5.10 and Eq. (5.59) may be re-investigated with the aid of vortex models presented above [5] or with full 3D CFD models ([13, 33, 34] in Chap. 5). As a result a new model for BEM applications could be formulated but still today there is few agreements on which tip-correction gives most accurate results.

7.7.2.1 Actuator Line Method

As a further example of the usefulness of the method, we mention the application of an extension, the *actuator line* [44] method to the MEXICO Sect. 8.6 experiment. Here, the forces are distributed on rotating lines modeling the blades. This usually is done using Gaussian smearing, and as a consequence a new parameter ε is introduced:

$$\eta_\varepsilon(x, y, z) = \frac{1}{\varepsilon^3 \pi^{3/2}} \cdot exp(-(x^2 + y^2 + z^2)/\varepsilon^2) . \tag{7.72}$$

Consistent determination of this parameter—which unfortunately influences the accuracy of the method considerably—is not as easy as originally thought. See [21] for further details

It must be noted that both Actuator disks and Lines rely on aerodynamic polars. With some justification AD/AL might be regarded as an extension of BEM. Therefore its shortcomings as how to formulate tip-correction approaches, for example remain.

7.8 Full-Scale CFD Modeling of Wind Turbine Rotors

In this section we discuss some of the results reached so far on applying CFD[2] to full 3D wind turbine rotor flow.

---

[2]From now on in this book the term CFD is assumed to be a RANS simulation where only the full 3D geometry of the wind turbine (or its blades) is used.

7.8.1 NREL Phase VI Turbine

One of the first successful applications of CFD was for the so-called NASA-AMES blind comparison (see Sect. 8.5) around the year 2000. Here, a very special rotor was chosen with two blades and a difficult S809 Profile. The main objective was to collect detailed load data from defined operating conditions in a wind tunnel, different and complementary to those gained in the free-field ([48] in Chap. 8). Figure 7.18 shows the shape and Fig. 7.19 measured polar data for Re $= 2 \cdot 10^6$ (Fig. 7.19). In Figs. 7.20 and 7.21 two simulations including transition modeling are shown. However, the first simulation (using a tabulated e^N method for transition prediction) is not able to capture c_D^{max}. Only one profile was used for the whole rotor so it was of utmost importance to capture even the 2D deep stall characteristics of this profile for power (or low shaft torque) prediction which is shown in Figs. 7.22 and 7.23. The results from Fig. 7.22 show that including modeling transition produces a correct trend even in the highly separated case ($v_{wind} > 10$ m/s). The same was done by Sørensen [47] including typical turbulence intensities (0.5 ... 1.5%) reported in Chap. 8 [59]. A remarkably strong dependence on turbulence intensity, especially in the velocity range around 11 m/s, is clearly seen. This emphasizes again that accurate transition modeling is absolutely necessary to accurately predict and understand wind turbine flow on wind turbine blades.

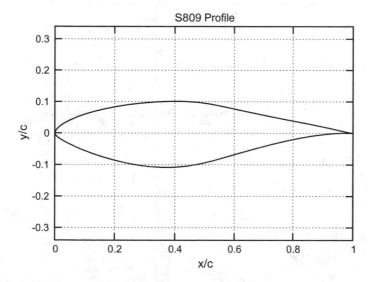

Fig. 7.18 Cross section of S809 airfoil used during the NASA-Ames blind comparison

7.8.2 The Mexico Rotor

During the IEAwind Annex 20 (2003–2008) the NASA-Ames unsteady experiment and blind comparison data was extensively investigated. In 2006 (see Sect. 8.6, [4]) and ([28] in Chap. 5) a European-Israel consortium constructed and measured a 3-bladed, 5 m diameter wind turbine in the European LLF wind tunnel (see Fig. 8.22). Shortly after another IEAwind Annex 29, *MexNext* was established. It was divided into two phases (01.06.2008–01.06.2011 and 1.11.2012–31.12.2014). Extensive CFD (not only one or two as in the NREL unsteady experiment, phase VI) simulation from at least 7 teams was used to extract useful information from the experiment. The blade was not equipped with only one profile, but 3 (DU91, Risoe-121 and NACA-64-4-18). See Figs. 7.24 and 7.25. A typical outcome is shown in Figs. 7.26 and 7.28 (from [29] in Chap. 5). Again the results from Risoe-DTU are closest to the measurements. Having had typical mesh sizes of several millions [46] in connection with the IEAwind Annex 20 CFD modeling the new mesh sizes grew considerably during the ensuing 10 years from IEAwind Annex 20 to 29: Figs. 7.29 and 7.29 [14] show the Chimera mesh topology and an typical vortical structure for TSR = 6.7 m/s and $v_{in} = 15$ m/s. The tau-Code from DLR, the German Aerospace Association was

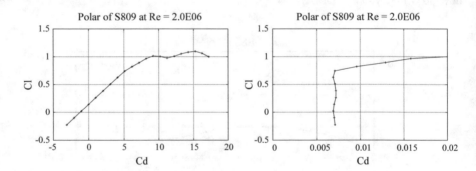

Fig. 7.19 S809: measured polar [39] ([49] in Chap. 8)

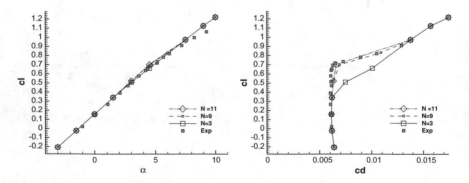

Fig. 7.20 S809: polar for $Re = 3 \cdot 10^6$, $Ma = 0.1$, [57] ([12] in Chap. 8)

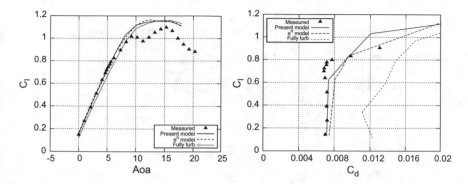

Fig. 7.21   S809: CFD simulation of [47]

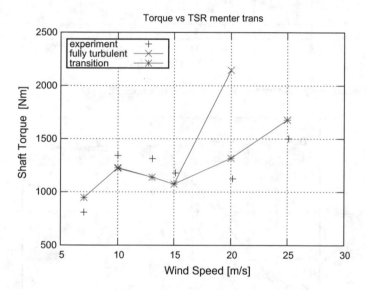

Fig. 7.22   NASA-Ames experiment. Torque versus wind speed, measurements, fully turbulent and transitional flow modeling, [17, 24]

used. It consists of 150 M cells and used the e^N transition modeling [27]. A typical job takes about 50 h on a 64-node parallel machine. Recently [6] reported a transient wake calculation for the MEXICO rotor with almost 1 G cells (850 M) with emphasis on investigations OF the tip-vortex breakdown phenomenon (Fig. 7.30).

As a last example we show in Fig. 7.31 our full set of applied computational methods to the MEXICO experiment. It shows the same amount of scatter AS in Figs. 7.26 and 7.28. Again in the attached flow case ($v < 15$ m/s inflow velocity) agreement for thrust (torque) is in the order of 10 (30)% only. It is further increased

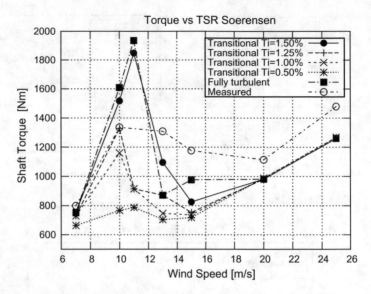

Fig. 7.23 Same as Fig. 7.22 but with different turbulence intensities [47]

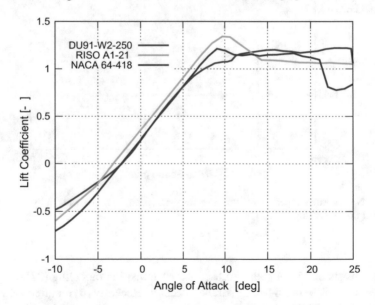

Fig. 7.24 Lift data of profiles: Cl versus AOA

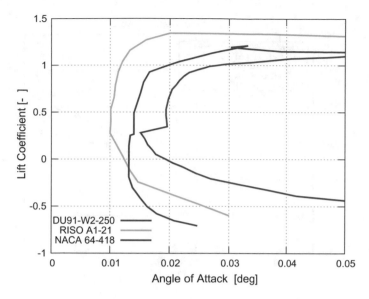

Fig. 7.25 Lift data of profiles: Cd versus Cl

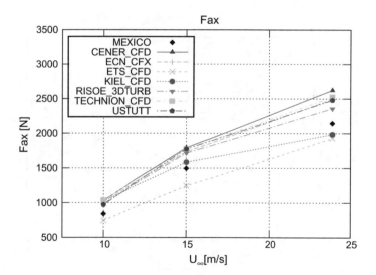

Fig. 7.26 Summary CFD (2014): thrust (axial force), from [29] in Chap. 5

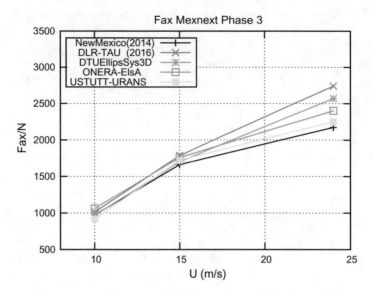

Fig. 7.27 Summary CFD (2019): thrust (axial force), from [29] in Chap. 5

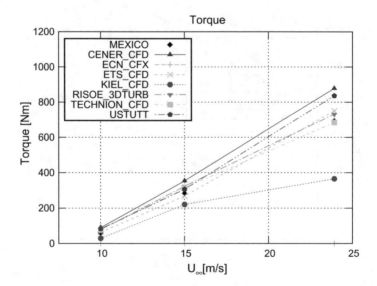

Fig. 7.28 Summary CFD: torque, from [29] in Chap. 5

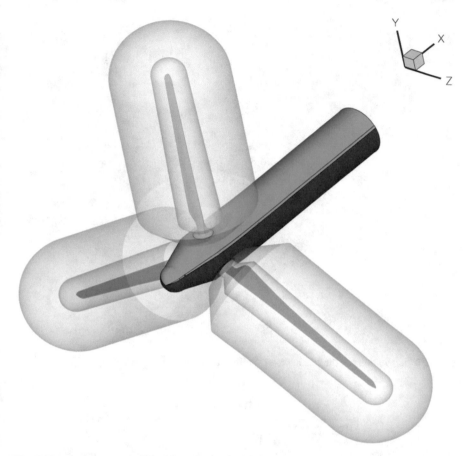

Fig. 7.29 Block structure of a 150 M cell mesh using chimera technique

when stall occurs (v > 15 m/s). An argumentation how to explain these findings could be conducted as follows:

The experiments should be accurate in a range of 3–5% [41]. BEM (light blue) use only the airfoil data given from 2D wind tunnel measurements. It should be accurate in the attached flow regime. Actuator Disk models with Fluent (red) and OpenFOAM (green) as flow solver use 2D airfoil data as well but the flow field is more complete. In principle this should reduce the error or at least the difference between measurement and BEM. Full 3D CFD (tau-code, blue) uses only the geometry of the blade and no airfoil data at all. However, as could be seen life is not that easy.

The experiment was repeated in 2014 [4] and the results were evaluated in IEAwind Annex 29 (MexNext) Phase 2 (2012–2014) [36] and Phase 3 (2015–2015) [2]. Some of the newer results are incorporated in Fig. 7.13 and some are shown in Fig. 7.27.

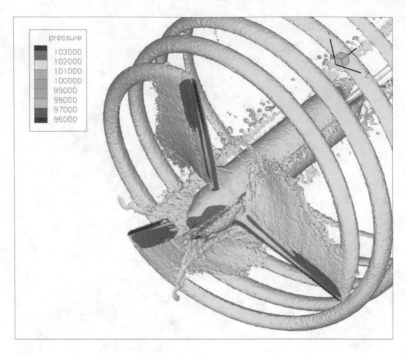

Fig. 7.30 Iso vorticity of the Mexico rotor for TSR of 6.7

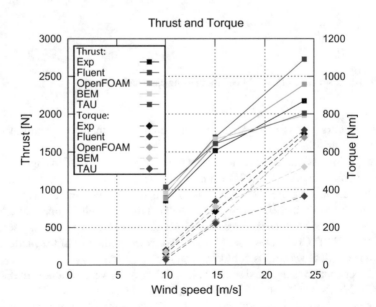

Fig. 7.31 Overview of some CFD methods applied to the Mexico experiment

7.8.3 *Commercial Wind Turbines*

We now enter a discussion about the use of CFD for commercial blades. As was stated several times, in the beginning of the first boom of wind turbine use (around 2000) the only reliable tool was BEM, sometimes with adapted *3D-corrected* airfoil data. Therefore it was tempting to apply CFD to these blades.

7.8.3.1 ARA48 Blades

This blade was designed and developed by aerodyn Energiesysteme, Germany together with Abeking & Rassmussen Rotoc. It was used in a 600/750 German wind turbine from *Huseumer Schiffswerft.* Later on other companies, Jacobs Wind-turbinen, Heide (now Senvion S.A.) and Goldwind, Beijing, China, manufactured several thousand of this sub-Megawatt turbine [34, 40]. Later the ARA blade family was complemented and substituted by LM blades. Figure 7.32 shows two different BEM designs as well as a measured curve. More about standards and how to measure a power curve will be seen in Sect. 8.2. An important conclusion was that CFD could only predict more accurate aerodynamic properties of stall blades if the turbulence model was able to do so. Clearly the model used by [34] (k-ε model) was not appropriate [10].

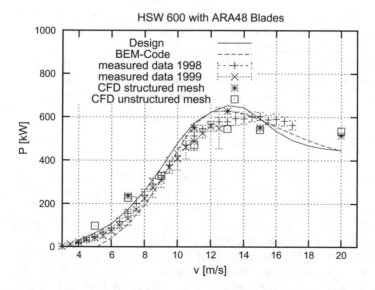

Fig. 7.32 Computed and measured power curve for a 400 kW stall machine, HSW 600 [34]

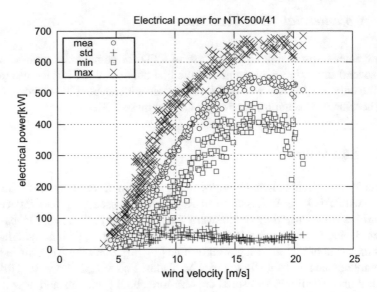

Fig. 7.33 Computed and measured power curve for a 400 kW stall machine, NTK 500 ([26] in Chap. 3)

7.8.3.2   NTK500

A comparable task for the Nortank NTK500 (stall) turbine [38] was undertaken somewhat earlier with aid of the actuator disk model [11] and without 2D airfoil data [45] by the Danish group, see Fig. 7.33.

As the machine (and blade) was designed using flow stall to limit output power, prediction of P_{max} was most important. Shortly after the NASA-Ames experimental data Sect. 8.5 took over becoming the model turbine for CFD testing.

7.8.3.3   Siemens Wind Turbine

CFD found its way to designers and manufacturers of wind turbines only very slowly. Main cause was the increasing number of so-called load cases (special conditions for inflow according to design rules) which demand very fast codes. Only BEM in its pure form is able to do so. In addition due to strong competitions results were published very rarely. One remarkable exception was information disseminated at the second *torque* conference [18]. A summary of results from simulation of a commercial 2.3 MW (93 m rotor diameter) is given in Table 7.2. ANSYS (CFX) is a commercial CFD code and Ellipsis is the research code developed by Risø-DTU. The deviations seem to be on the order of few percentages only. Remarkably, as also seen in the cases discussed above, transition modeling does not improve accuracy in all cases. Clearly much more comparisons have to be undertaken. A somewhat similar comparison (using a different turbine) and using ANSYS-CFX and OpenFOAM was performed

Table 7.2 Mechanical power (kW) prediction for a commercial (SWT-2.3-93) wind turbine, from [18]

Wind (m/s)	RPM	Measured	ANSYS, fully turb	ANSYS, transitional	Ellipsis	BEM-code
6	10.5	400	396	428	392	408
8	13.5	986	950	–	945	967
10	16.0	1894	1853	1977	1850	1850
11	16.0	2323	2388	–	–	–

by Dose [7]. His findings agree with those from [18], with the exception that an OpenFOAM run needs much more CPU time (more than a factor of 10) so far.

7.8.3.4 NM80 Wind Turbine—IEAwind Task 29, Phase 4

After phase 3 (2012–2017) IEAwind Task 29 was renamed to *Analysis of Aerodynamic Measurements* and extended to a duration from 2018 to 2020. Based on the DAN-AERO experiment ([21–23] in Chap. 8) measured data was made available to the participants and BEM as well as CFD calculations were caried out. The turbine has a rotor diameter of 80 m and blades from LM (LM38.8) were used. Figure 7.34 shows a preliminary comparison from CFD Codes (first seven columns), BEM (next), and an estimation of measured power (last red box). Agreement between the first six codes is within few percent. Results from column 7 are from a 3D panel code FUNAERO [3].

7.9 Concluding Remarks About Use (and Abuse) of CFD

This chapter could not be finished without some remarks on the state of the art and future progress on CFD methods applied to wind turbines. Clearly CFD—defined as numerical solution of the NAST including some turbulence modeling—is the only way to improve ordinary BEM. As reviewed in [52], two equations models have a long history, only slightly surpassed in recent times by LES and hybrid RANS/LES models [51]. So by now the accuracy of any CFD calculation is limited by the (sometimes accidental) accuracy of the turbulence model. As explained in Chap. 5 many researchers agree ([36] in Chap. 5) that only a deeper understanding of the nature of turbulence and the mathematical properties of the Navier–Stokes Equations will lead to controlled improved turbulence models for use in wind turbine aerodynamic applications.

The only independent way for assessment is the use of experiments which will be described and discussed in the following Chap. 8.

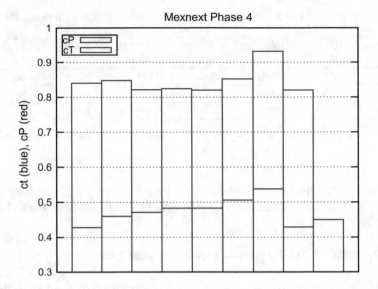

Fig. 7.34 Results form a comparison of CFD calculation for the NM80 rotor. Inflow is 6.1 m/s and RPM = 12.3

7.10 Problems

Problem 7.1 Turbulent boundary layers for $20 < y^+ < 200$ show a logarithmic velocity profile:

$$u^+ = \frac{1}{\chi} \cdot ln(y^+) + 5.5, \quad \chi = 0.41 \tag{7.73}$$

Then k-ε—simplifies to

$$-\partial_y \left(c_\mu \frac{k^2}{\varepsilon} \frac{\partial k}{\partial y} \right) - c_\mu \frac{k^2}{\varepsilon} E + \varepsilon = 0 , \tag{7.74}$$

$$-\partial_y \left(c_\varepsilon \frac{k^2}{\varepsilon} \frac{\partial k}{\partial y} \right) - c_1 k E + c_2 \frac{\varepsilon^2}{k} = 0 . \tag{7.75}$$

Show that with

$$k(y) \quad = \frac{u^{*2}}{\sqrt{c_\mu}} , \tag{7.76}$$

$$\varepsilon(y) = \frac{u^{*3}}{\sqrt{\chi y}} \text{ and} \tag{7.77}$$

$$E(y) \quad = \frac{u^{*2}}{\chi^2 y^2} \tag{7.78}$$

the above equations may be solved only if

$$c_\varepsilon = \frac{\sqrt{c_\mu}}{\chi^2} \left(c_2 c_\mu - c_1\right) (= 0.08) \,. \tag{7.79}$$

Remark: $u^+ := U(y)/u^*$, $y^+ := y/y^*$, $y^* = v/u^*$, $u^* = v = const$

Problem 7.2 Decaying homogeneous turbulence.

In case $\mathbf{u}$ and $\nabla \mathbf{u}$ both are zero, the model k-ε-equations reduce to

$$\partial_t k + \varepsilon = 0 \,, \tag{7.80}$$

$$\partial_t \varepsilon + c_2 \frac{\varepsilon^2}{k} = 0 \,. \tag{7.81}$$

Show that a polynomial decay in time:

$$k(t) = k_0 \cdot (1 + \lambda t)^{-n} \,, \tag{7.82}$$

$$\varepsilon(t) = \varepsilon_0 \cdot (1 + \lambda t)^{-m} \,, \tag{7.83}$$

obeys the equation. Use n $= 1.3$ from experiments to find a value for c_2

Problem 7.3 Falkner–Skan Flow with simple Fortran Code. At the edge of the boundary layer $u_e = u_0 \cdot (x/L)^m$. Define $\beta := 2m/(1+m)$. With

$$\eta = y\sqrt{\frac{u_0(m+1)}{2vL}} \left(\frac{x}{L}\right)^{\frac{m-1}{2}} \tag{7.84}$$

the BL DEQ reads

$$f''' + f \cdot f'' + \beta \left(1 - f'^2\right) = 0 \,. \tag{7.85}$$

Use the simple code ([64] in Chap. 3) from Fig. 7.35 for solving (Table 7.3).

Problem 7.4 DU-W-300-mod

(a) Install XFoil or JavaFoil on your computer. For the profile of Fig. 7.19:
(b) Extract a polar: c_L versus c_D for Re $= 1.15, 3.6$, and $10 \cdot 10^6$
(c) Compare to measured data of Figs. A.31 and A.32.

Problem 7.5 (a) Install a CFD (RANS) code on your computer. For the same profile
(b) extract a polar (c_L vs. c_D) for Re $= 1.2, 5.8$, and $10.2 \cdot 10^6$
(c) Compare to measured data of Figs. A.31 and A.32.

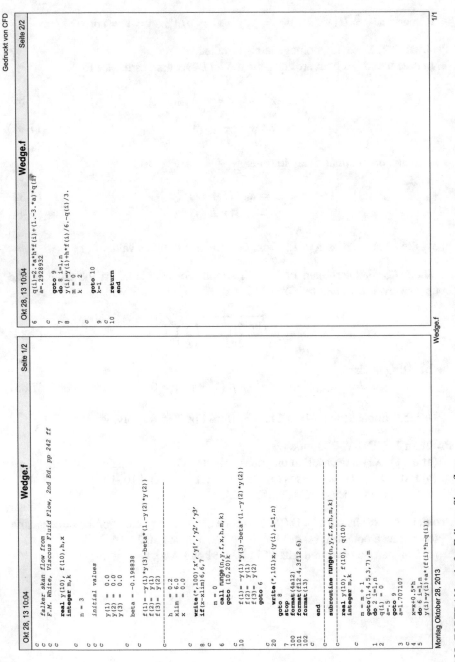

Fig. 7.35 FORTRAN code for Falkner–Skan flow

Table 7.3   Coordinates for DU-W-300-mod

x/c	y/c	x/c	y/c	x/c	y/c	x/c	y/c
1.000000	0.000000	0.159531	0.125199	0.001989	−0.016639	0.279441	−0.165530
0.991346	0.002458	0.150115	0.123207	0.002724	−0.019462	0.290216	−0.164664
0.977819	0.006263	0.141097	0.121057	0.003584	−0.022332	0.301248	−0.163440
0.963559	0.010090	0.132472	0.118773	0.004565	−0.025252	0.312578	−0.161850
0.948539	0.013944	0.124231	0.116351	0.005673	−0.028222	0.324247	−0.159870
0.932787	0.017867	0.116366	0.113822	0.006920	−0.031240	0.336318	−0.157508
0.916224	0.021898	0.108870	0.111183	0.008300	−0.034310	0.348825	−0.154758
0.898905	0.026052	0.101728	0.108456	0.009817	−0.037435	0.361820	−0.151588
0.880975	0.030297	0.094937	0.105634	0.011477	−0.040615	0.375355	−0.148006
0.862597	0.034583	0.088475	0.102750	0.013287	−0.043851	0.389491	−0.143981
0.843925	0.038881	0.082343	0.099782	0.015248	−0.047149	0.404275	−0.139526
0.824935	0.043199	0.076513	0.096773	0.017361	−0.050511	0.419747	−0.134619
0.805770	0.047494	0.070988	0.093707	0.019636	−0.053939	0.435927	−0.129310
0.786591	0.051727	0.065750	0.090603	0.022078	−0.057432	0.452772	−0.123594
0.767284	0.055925	0.060785	0.087473	0.024701	−0.060990	0.470261	−0.117492
0.747636	0.060133	0.056091	0.084311	0.027514	−0.064609	0.488327	−0.111078
0.727658	0.064355	0.051649	0.081134	0.030525	−0.068291	0.506850	−0.104353
0.707496	0.068545	0.047455	0.077944	0.033739	−0.072040	0.525745	−0.097433
0.687207	0.072701	0.043491	0.074750	0.037163	−0.075853	0.544833	−0.090427
0.666950	0.076790	0.039759	0.071550	0.040815	−0.079720	0.564031	−0.083357
0.646893	0.080764	0.036237	0.068361	0.044700	−0.083639	0.583174	−0.076325
0.627015	0.084639	0.032925	0.065176	0.048834	−0.087601	0.602089	−0.069419
0.607140	0.088433	0.029805	0.062010	0.053224	−0.091602	0.620884	−0.062634
0.587170	0.092169	0.026875	0.058862	0.057878	−0.095632	0.639535	−0.055977
0.567157	0.095824	0.024128	0.055734	0.062809	−0.099682	0.658011	−0.049507
0.547197	0.099394	0.021557	0.052627	0.068024	−0.103741	0.676178	−0.043263
0.527321	0.102858	0.019155	0.049543	0.073527	−0.107800	0.693952	−0.037306
0.507499	0.106222	0.016908	0.046491	0.079329	−0.111843	0.711285	−0.031676
0.487780	0.109452	0.014820	0.043464	0.085432	−0.115858	0.728146	−0.026381
0.468108	0.112505	0.012876	0.040472	0.091832	−0.119839	0.744508	−0.021448
0.448540	0.115431	0.011080	0.037507	0.098528	−0.123764	0.760365	−0.016889
0.429194	0.118206	0.009427	0.034572	0.105522	−0.127617	0.775738	−0.012722
0.410155	0.120805	0.007911	0.031668	0.112804	−0.131382	0.790647	−0.008936
0.391510	0.123195	0.006526	0.028797	0.120371	−0.135039	0.805139	−0.005536
0.373327	0.125381	0.005275	0.025956	0.128212	−0.138565	0.819249	−0.002527
0.355667	0.127331	0.004160	0.023144	0.136323	−0.141944	0.833003	0.000131
0.338577	0.129045	0.003180	0.020361	0.144689	−0.145162	0.846429	0.002425
0.322114	0.130510	0.002330	0.017608	0.153290	−0.148215	0.859546	0.004367

(continued)

Table 7.3   (continued)

x/c	y/c	x/c	y/c	x/c	y/c	x/c	y/c
0.306299	0.131712	0.001625	0.014883	0.162111	−0.151081	0.872385	0.005956
0.291143	0.132654	0.001066	0.012187	0.171136	−0.153743	0.884978	0.007192
0.276635	0.133321	0.000640	0.009524	0.180335	−0.156192	0.897349	0.008098
0.262759	0.133722	0.000331	0.006899	0.189687	−0.158416	0.909509	0.008675
0.249473	0.133830	0.000134	0.004315	0.199181	−0.160382	0.921472	0.008933
0.236726	0.133641	0.000026	0.001774	0.208793	−0.162099	0.933261	0.008858
0.224457	0.133149	0.000005	−0.000720	0.218516	−0.163537	0.944897	0.008433
0.212625	0.132392	0.000081	−0.003244	0.228346	−0.164679	0.956443	0.007607
0.201202	0.131376	0.000249	−0.005824	0.238282	−0.165522	0.967976	0.006337
0.190187	0.130133	0.000513	−0.008454	0.248333	−0.166030	0.979640	0.004523
0.179567	0.128679	0.000887	−0.011135	0.258532	−0.166201	0.991639	0.002067
0.169349	0.127025	0.001376	−0.013863	0.268893	−0.166045	1.000000	0.000000

References

1. Anderson TJD Jr (1995) Computational fluid dynamics. McGraw-Hill, New York
2. Boorsma, K, Schepers JG, Gomez-Iradi S, Herraez I, Lutz T, Weihing P, Oggiano L, Pirrung G, Madsen HA, Shen WZ, Rahimi H, Schaffarczyk AP (2018) Final report of IEA Wind Taks 29 Mexnext (phase 3), ECN-E–18-003, Petten, The Netherlands
3. Boorsma K, Greco L, Bedon G (2018) Rotor wake engineering models for aeroelastic applications. In: Proceeding of TORQUE2018. IOP Conf Ser: J Phys: Conf Ser: 1037:062013
4. Boorsma K, Schepers G (2014) New Mexico experiment, ECN-E–14-048, Petten, The Netherlands
5. Branlard E (2013) Wind turbine tip-loss correction. Master's Thesis (public version). Risø DTU, Copenhagen, Denmark
6. Carrion M (2013) Understanding wind turbine wake breakdown using CFD, 3rd. IEA wind MexNext II Meeting, Pamplona, Spain
7. Dose B (2013) CFD simulations of a 2.5 MW wind turbine using ANSYS CFX and OpenFOAM. MSc Thesis, UAS Kiel and FhG IWES, Germany
8. Ferziger JH, Perić M (2002) Computational methods for fluid dynamics, 3rd edn. Springer, Berlin
9. Fletcher CAJ (2005) Computational techniques for fluid dynamics, 2 volumes. Springer, Berlin
10. Hansen MOL (1998) Private communication
11. Hansen MOL et al (1997) A global Navier-Stokes rotor prediction model. AIAA 97–0970, Reno
12. Ishihara T, Gotoh T, Kaneda Y (2009) Study of high-reynolds number isotropic turbulence by direct numerical simulation. Annu Rev Fluid Mech 41:165–180
13. Jeromin A, Bentamy A, Schaffarczyk AP (2013) Actuator disk modeling of the Mexico rotor with openFOAM. In: 1st Symposium on openFOAM in wind energy, Oldenburg, Germany
14. Jeromin A, Schaffarcyzk AP (2012) First steps in simulating laminar-turbulent transition on the Mexico blades, 2nd. IEAwind MexNext II Meeting, NREL, Golden, CO, USA
15. Kolmogorov AN (1942) Equations of turbulent motion of an incompressible fluid. Izv Akad Nauk SSSR, Ser Fiz VI(1-2):56–58
16. La Yi et al (2008) A public turbulence database cluster and applications to study Lagrangian evolution of velocity increments in turbulence. J Turb 9(31):1–20

17. Langtry RB (2006) A correlation-based transition model using local variables for unstructured parallelized CFD codes. Dissertation, Universität Stuttgart
18. Laursen J, Enevoldsen P, Hjort S (2007) 3D CFD quantification of the performance of a multi-megawatt wind turbine. In: Proceedings of 2nd conference of the science of making torque from wind, Copenhagen, Denmark
19. Lobo BA, Boorsma K, Schaffarczyk AP (2018) Investigation into boundary layer transition on the Mexico blade. Proceeding of TORQUE2018. IOP Conf Ser: J Phys: Conf Ser: 1037:052020
20. Mahmoodi E, Schaffarczyk AP (2012) Actuator disc modeling of the Mexico rotor experiment. In: Proceeding of Euromech Coll. 528, Wind energy and the impact of turbulence on the conversion process, Oldenburg, Germany
21. Martinez-Tossa LA, Meneveau C (2019) Filtered lifting line theory and application to the actuator line method. J Fluid Mech 863:269–292
22. Menter F (1992) Improved two-equation k-ω turbulence models for aerodynamical flows. NASA Technical Memorandum 103975. Moffett Field, CA, USA
23. Menter F (1994) Two-equation eddy-viscosity turbulence models for engineering applications. AIAA - J 32(8):1598–1605
24. Menter F, Langtry R (2006) Transitionsmodellierungen technischer Strömungen (Modeling of transitions in engineering flow). ANSYS Germany, Otterfing, Germany
25. Michelassi V, Rodi W, Zhu J (1993) Testing a low Reynolds number k-ε model based on direct simulation data. AIAA - J 31(9):1720–1723
26. Mohammandi B, Pirronneau O (1994) Analysis of the k-epsilon turbulence model. Wiley, Chichester
27. NN (2009) Transition module (V8.76) user guide (V1.0 beta). Unpublished report, Braunschweig, Germany (in German)
28. OpenFOAM Foundation (2013) User guide: openFOAM, the open source CFD toolbox
29. Prandtl L (1925) Bericht über Untersuchungen zur ausgebildeten Turbulenz. Z angew Math und Mech 5
30. Prandtl L (1945) Über ein neues Formelsystem für die ausgebildete Turbulenz. Nachr. d. Akad. d. Wiss. in Göttingen, Math.-nat. Klasse, S. 6–20
31. Reichstein T, Schaffarczyk AP, Dollinger C, Balaresque N, Schülein E, Jauch C, Fischer A (2019) Investigation of laminar-turbulent transition on a rotating wind-turbine blade of multi-megawatt class with thermography and microphone array. Energies 12(11):2102
32. Sanders B, van der Pijl SP, Koren B (2011) Review of computational fluid dynamics for wind turbine wake aerodynamics. Wind Energy 14:799–819
33. Schaffarczyk AP, Boisard R, Boorsma K, Dose B, Lienard C, Lutz T, Madsen HA, Rahimi H, Reichstein T, Schepers G, Sørensen N, Stoevesand B, Weihig P (2018) Comparison of 3d transitional CFD simulations for rotating wind turbine wings with measurements. In: Proceeding of TORQUE2018. IOP Conf Ser: J Phys: Conf Ser: 1027:022012
34. Schaffarczyk AP (1997) Numerical and theoretical investigation for wind turbines. IEAwind, Annex XI meeting, ECN, Petten, The Netherlands
35. Schaffarczyk AP, Schwab D, Breuer M (2016) Experimental detection of laminar-turbulent transition on a rotating wind turbine blade in the free atmosphere. Wind Energy 20(2)
36. Schepers JG, Boorsma K, Gomez-Iradi S, Schaffarczyk AP, Madsen HA, Sørensen NN, Shen WZ, Lutz T, Schulz C, Herraez I, Schreck S (2014) Final report of IEA Wind Taks 29 Mexnext (phase 2), ECN-E–14-060, Petten, The Netherlands
37. Schlatter P (2005) Large-eddy simulation of transitional turbulence in wall-bounded shear flow. PhD thesis ETH, No. 16000, Zürich, Switzerland
38. Schmidt Paulsen U (1995) Konceptundersøgelse Nordtanks 500/41 Strukturelle laster, Risø-I-936(DA), Roskilde, Denmark
39. Schreck S (2008) IEA wind annex XX: Hawt aerodynamics and models from wind tunnel measurements. NREL/TP-500-43508. Golden, CO, USA
40. Schubert M, Schumacher K (1996) Entwurf einer neuen Aktiv-Stall Rotorblattfamilie (Design of a new family of active stall blades). In: Proceeding of DEWEK '96, Wilhelmshaven, Germany (in German)

41. Shen WZ, Zhu WJ, Sørensen JN (2012) Actuator line/Navier-Stokes computations for the Mexico rotor: comparison with detailed measurements. Wind Energy 15(5):811–825
42. Smith LM, Reynolds WC (1992) On the Yakhot-Orzag renormalization group method for deriving turbulence statistics and models. Phys Fluids A 4:364
43. Smith LM, Reynolds WC (1998) Renormalization group analysis of turbulence. Ann Rev Fluid Mech 30:275–310
44. Sørensen JN, Shen WZ (2002) Numerical modelling of wind turbine wakes. J Fluids Eng 124(2):393–399
45. Sørensen NN, Hansen MOL (1998) Rotor performance prediction using a Navier-Stokes method. AIAA-98-0025. Reno, NV, USA
46. Sørensen NN, Michelsen JA, Schreck S (2002) Navier-Stokes prediction of the NREL phase VI rotor in the NASA Ames 80 ft x 120 ft wind tunnel. Wind Energy 5:151–168
47. Sørensen NN (2009) CFD modeling of laminar-turbulent transition for airfoils and rotors using the $\gamma - \tilde{R}e_\theta$ model. Wind Energy 12(8):715–733
48. Sørensen NN 1998) HypGrid2D a 2-D mesh generator, Risø-R-1035(EN)
49. Spalart P (1988) Direct simulation of a turbulent boundary layer up to $R_\theta = 1410$. J Fluids Eng 187:61–98
50. Spalart P (2000) Strategies for turbzulence modelling and simulations. Int J Heat Fluid Flow 21:225–263
51. Spalart P (2009) Detached-eddy simulation. Annu Rev Fluid Mech 41:181–202
52. Spalding B (1991) Kolmogorov's two-equation model of turbulence. Proc R Soc 434:211–216
53. Trede R (2003) Entwicklung eines Netzgenerators. Diplomarbeit, FH Westküste, Heide (in German)
54. Tu J, Yeoh GH, Liu C (2008) Computational fluid dynamics. Butterworth-Heinemann, Elsevier, Amsterdam
55. von Neumann J (1949) Recent theories of turbulence. Unpublished report to the Naval office, in collected work (S. Ulam, Ed.) VI:437–472
56. Wilcox DC (1993/94) Turbulence modeling for CFD. DCW Industries Inc
57. Winkler H, Schaffarczyk AP (2003) Numerische Simulation des Reynoldszahlverhaltens von dicken aerodynamischen Profilen für Off-shore Anwendungen. Bericht des Labors für numerische Mechanik, 33, Kiel (in German)
58. Wu YT, Port-Agel F (2013) Modeling turbine wakes and power losses within a wind farm using LES: an application to the Horns-Rev offshore wind farm. In: Proceeding of ICOWES 2013, Copenhagen, Denmark
59. Yakhot V, Orszag SA (1986) Renormalization group analysis of turbulence. I. Basic theory. J Sci Comput 1(1)

Chapter 8
Experiments

> *Everybody believes in measurements—except the experimentalist. Nobody believes in theory—except the theorist (Unknown source).*

This chapter describes methods and results which have been applied to gain insight into the physics of wind turbine flow by performing experiments.

8.1 Measurements of 2D Airfoil Data

As we have explained extensively in the previous chapters the blade-element-momentum method describes blades—aerodynamically—as a set of independent 2D airfoils. Therefore, wind tunnel measurement of 2D sections is the basis of all aerodynamical experiments for wind turbines. Fortunately, a lot of experience has been gained for airfoils of airplanes which could be used when special airfoils started to be designed [33] in Chap. 9. The requirements for a wind turbine blade airfoil is different in several respects:

- a high lift-to-drag ratio is most important,
- c_L^{max} should be limited to values not far from $c_L^{L2D-max}$
- thick ($>30\%$) airfoils have to be used at least in the inner part of the blade.

In Fig. 8.1 we show a profile and its setup for measurement campaigns using a 30% thick airfoil (slightly modified DU97-W-300 profile from Technical University Delft [55]).

Some results [12, 55] are summarized in Figs. 8.2, 8.3, and 8.4.

Comparing Figs. 8.2 and 8.3, the last performed at RN = 5 M, we see clearly the failure of XFoil (a popular open-source panel boundary layer code, see Chap. 3) to predict c_L^{max}.

c_L^{max} may only be predicted with appropriate turbulence models, e.g., Menter's [22] in Chap. 7.

© Springer Nature Switzerland AG 2020
A. P. Schaffarczyk, *Introduction to Wind Turbine Aerodynamics*,
Green Energy and Technology, https://doi.org/10.1007/978-3-030-41028-5_8

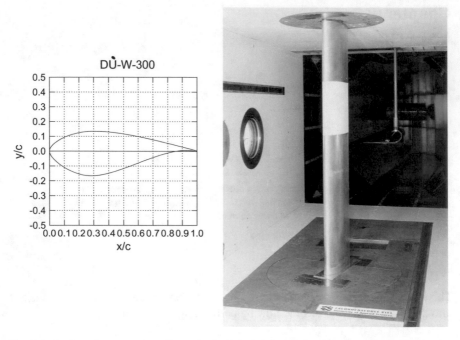

Fig. 8.1   Outline of measured profile DU97-W-300-mod and model setup in the wind tunnel

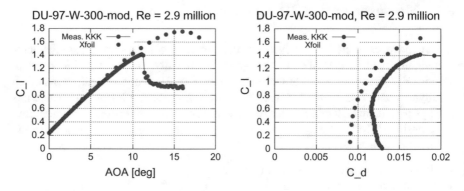

Fig. 8.2   DU97-W-300-mod measured at Kölner KK and XFoil, Re = 2.85 M

Usually, only global forces are measured by gauges and/or *wake rakes* only. A common but elaborate technique uses small holes (diameter $\emptyset < 0.5$ mm) on the surface which are connected to pressure transducers. Figure 8.4 shows one example of such a circumferentially resolved measurement. Small laminar separation bubbles or sometimes even the transition from laminar to turbulent flow (by the displacement effect) sometimes are visible (small bump at x/c = 0.2 for the XFoil data) in the c_p-distribution. Notice that the pressure does not indicate a stagnation point ($c_p = 1$) at the tail. Even thicker (from 35% up about 50%) [2, 34, 58] profiles are used

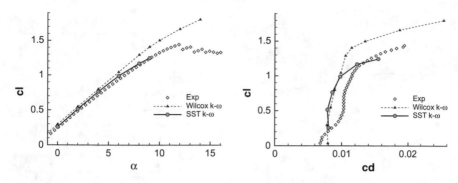

Fig. 8.3 DU97-300-mod simulated with FLOWer and SST k-ω turbulence model, Re $= 5$ M

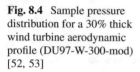

Fig. 8.4 Sample pressure distribution for a 30% thick wind turbine aerodynamic profile (DU97-W-300-mod) [52, 53]

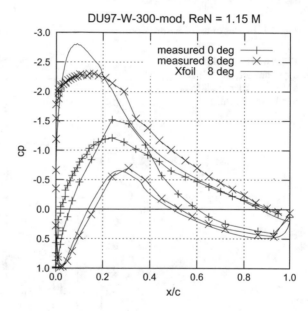

in the very inner part of a blade for improving structural integrity (see Chap. 9). Figures 8.5 and 8.6 show shapes of two 35 (50)% profiles. Similar comparisons have been performed at the LM wind tunnel by [20], see Fig. 8.7.

Summarizing the findings from [2, 20, 34, 58] we may conclude: Capture of c_L^{max} is somewhat easier by measurements but very difficult by CFD even when using DES. It has to be noted that a high (>100) lift-to-drag ratio is possible only at Reynolds number $>5 \cdot 10^5$ [17]. This has to be taken into account for an airfoil to be used for blades in small wind turbines and wind tunnel models.

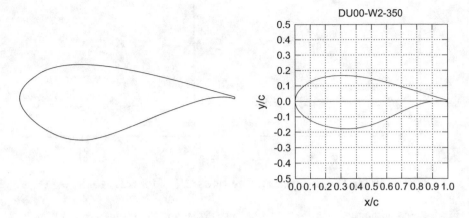

Fig. 8.5 Shapes of two 35% thick profiles: DU96-W-351 and DU00-W2-350

Fig. 8.6 Two examples for approximately 50% thick airfoils for root sections of wind turbine blades

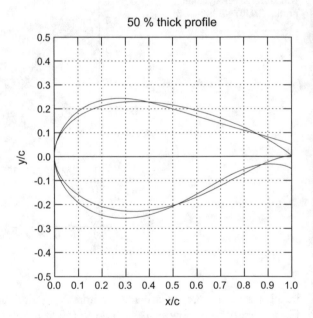

8.1.1   DU00-W-210

During the AVATAR project, a blind test was carried for DU00-W-210 which has been measured at DNW-HDG. Several CFD-codes as well as panel code were asked to investigate Re- as well as TI-dependence of lift- and drag data. Figure 8.8 shows one of the results. First of all, L2D data can be predicted with less than 3% (Ellipsys from DTU) accuracy. It seems that the e^N method is more accurate to predict transition than the correlation method. Finally, it is clearly seen that fully turbulent calculations

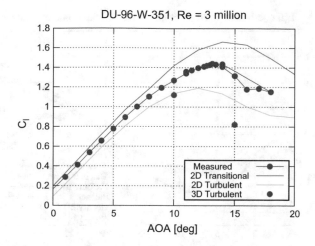

Fig. 8.7 Lift data for a 35% thick airfoil (DU96-W-351) including 3D CFD simulation from [20]

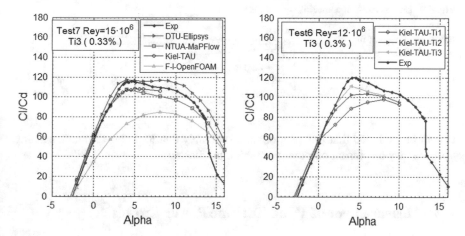

Fig. 8.8 CFD simulation for DU00-W-210 with transitional CFD-Codes. Left: Various CFD-Codes. Right: TI-dependence from [25]

underestimate max-L2D by more than 30%. TI-dependence (by Mack's correlation Eq. 8.3) can be modeled accurately as well.

8.1.2 Very Thick Airfoils

Figure 8.9 shows c_L versus AOA data of a special airfoil-type profile comparable to those of Fig. 8.6. As a unusual feature it exhibits regions of negative lift-slope. It

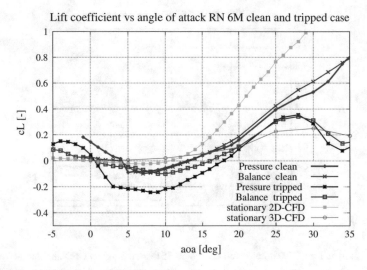

Fig. 8.9 Lift vs. AOA of a special, 46% thick profile

Fig. 8.10 Shape of the
ARA-48 blade

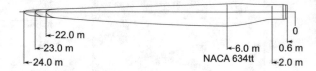

cannot be excluded that this phenomena, problably due to a pair of counter-acting
vortices may be present on other very thick profiles.

8.2 Measurement of Wind Turbine Power Curves

A reliably measured power curve of a wind turbine is of utmost importance for the
economic success of a wind turbine as a product. We already have seen in Chap. 7
two examples, see Figs. 7.32 and 7.33. Standards [14] exist how the terrain should be
classified and which measurement equipment has to be used when a *certified* power
curve $P(v)$ is desired. This is especially important when no (expensive) separate met
mast is available. Then the usual (special calibrated) nacelle anemometer is the only
device for measuring the wind speed. Figure 8.12 shows a sample measured power
curve for a small wind turbine (Figs. 8.10, 8.11 and 8.13).

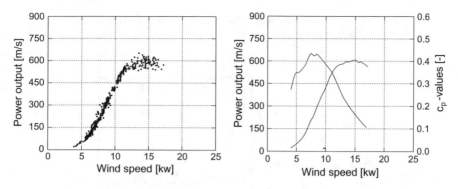

Fig. 8.11 Measured power of blade from Fig. 8.10. Left: Raw data. Right: Averaged power curve (right scale) and c_P (left scale) as function of wind speed

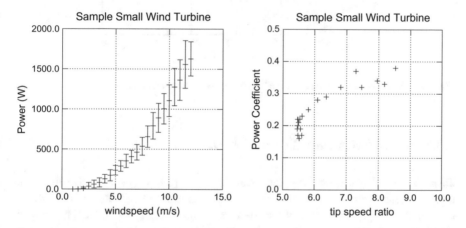

Fig. 8.12 Power versus wind speed (left) and c_P versus λ (right) for a small wind turbine

8.3 IEAwind Field Rotor Experiments

8.3.1 The International Energy Agency and IEAwind

The IEAwind is a branch of the International Energy Agency (IEA) and was founded comparatively early (in 1974) and defines its work as quoted here:

> The International Energy Agency (IEA) Wind agreement is a vehicle for member countries to exchange information on the planning and execution of national large-scale wind system projects and to undertake co-operative research and development (R & D) projects called Tasks or Annexes.

It is therefore a unique place for scientists from all around the world to communicate about wind turbine aerodynamics. First results were published in the early 1980 and one (Task XI *Base Technology Information Exchange*) is continuously working.

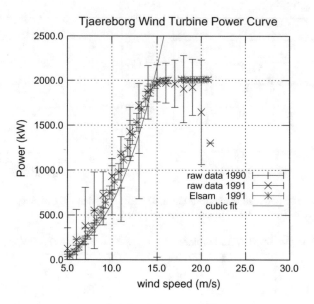

Fig. 8.13 Power versus wind speed for a 2 MW test wind turbine, based on 10 min averages data from [11]

8.3.2   IEAwind Annex XIV and XVIII

Task 14 and 18 (for more complete overviews see [33, 42]) summarize first field measurements for comparatively small (around 25 m rotor diameter) wind turbines and supported the reinvention of the so-called *Himmelskamp effect* [12] in Chap. 5. One important task was to compare the very different experimental approaches and thereby to identify the accuracy of the measurements (Figs. 8.14 and 8.15).

The Himmelskamp effect—in short—is an effect of delay of stall at high angles of attack. Much higher c_L values than expected may be reached when the rotor is in operation compared to the 2D case. Examples are shown in Figs. 8.16, 8.17 and 8.20.

These measurements are very time-consuming and expensive. One special point is the long averaging period due to the heavily transient inflow conditions. Therefore, wind tunnel measurements were planned. We will introduce this in more detail in Sects. 8.4 and 8.5.

8.3.3   Angle of Attack in 3D Configuration

Questions arising from instrumentation were discussed in Sect. 8.3 as well as the important question how to relate these data to the 2D polars used in BEM. One special point is to convert measured normal (usually in wind direction) and tangential

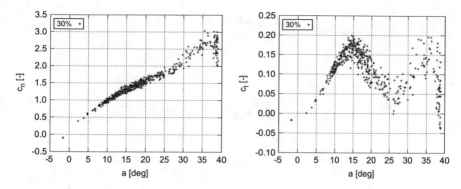

Fig. 8.14 Raw data for normal and tangential for r/R = 30%

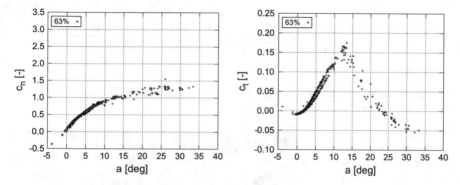

Fig. 8.15 Raw data for normal and tangential force r/R = 63%

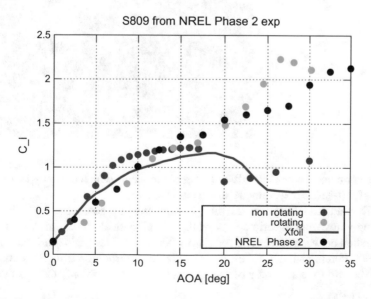

Fig. 8.16 Example of stall-delay on a rotating blade measured in the field, data from [37, 39]

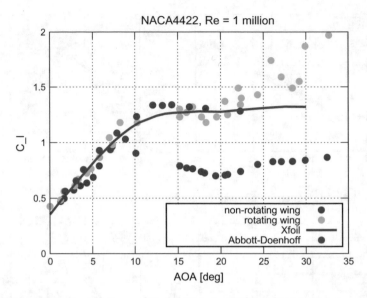

Fig. 8.17 Exampled of stall-delay on a rotating blade from the FFA/CARDC wind tunnel experiment. Data from [7]

Fig. 8.18 Outline of the NASA-AMES wind tunnel, the open inlet has dimensions of $108 \times 39\,\text{m}^2$ size, *source* NASA

(in the rotor plane) forces into lift and drag components which can only be done if a meaningful (total) inflow velocity direction is known.

In 2D wind tunnel experiment this is guaranteed by geometric conditions. For a rotating wing in an experiment as well as in numerical model, there is a strong inter-relation between forces and local flow because of Newton's third axiom.

Many different methods are used to extract this information from force measurements (see [46] for a summary of recent results). To mention only the most used

- inverse BEM,
- Biot–Savart,

Fig. 8.19 Outline of the NASA-AMES (NREL Phase VI) 2-blades rotor experiment (the cross section for testing is 24.4 × 36.6 m². With permission of NWTC/NREL, Golden, CO, USA

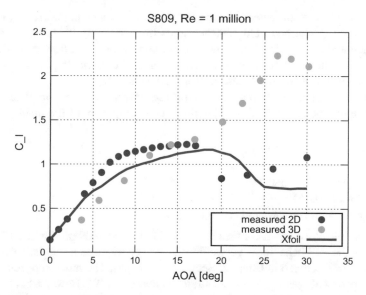

Fig. 8.20 Exampled of stall-delay on a rotating blade from the NASA-Ames Phase VI experiment, data from [57]

- stagnation point method and
- direct measurement of the velocity vector.

So far it can be said that no simple solution is possible and one should use more than one method if one wants to extract quasi 2D data from 3D configurations.

8.4   FFA/CARDC Wind Tunnel Experiments

One of the first wind tunnel measurements was conducted in a Chinese–Swedish co-operation during the years 1986 and 1992 [7, 9, 31]. The blade had a radius of 2.648 m and was equipped with NACA 44tt [1] with $tt = 14$ to 22%. Reynolds number in general was about 500 k. A main finding was the confirmation of the stall-delay at sections closer the hub, see Fig. 8.17.

The difference is particularly large when comparing the rotating and non-rotating case, but the stall behavior seems to be very different from what is expected by older measurements [1] and 2D Boundary Layer Panel codes [14] in Chap. 3.[1]

8.5   NREL NASA-Ames Wind Tunnel Experiment

The *National Renewable Energy Laboratory (NREL)*, located in the vicinity of Denver, Colorado, performed field measurements from the early 90ies on [8, 48]. As mentioned earlier, the idea was to have more control on the inflow conditions, so a larger wind tunnel campaign was performed in the world largest wind tunnel ($24.4 \times 36.6\,m^2$), at NASA-Ames, see Fig. 8.19 (Fig. 8.18).

The blade was designed from only one profile, the S809 [51]. A complete description of the turbine is included in [3]. Other main parameters of the machine are as follows:

- rotor diameter 10.0–11.06 m,
- hub height 12.2 m,
- cone angle: $0°$, $3.4°$ and $18°$,
- rated power 19.8 kW,
- 2-bladed up- and down-wind configuration.

As an example of the again confirmed stall-delay for the S809 profile, see Fig. 8.20. Predicting the stall behavior is very complicated and required a sophisticated CFD-model as was discussed in Chap. 7, Sect. 7.8.1 in detail. The most important aspect of the experiment was the so-called *blind comparison* [49]. Here participants had to calculate power and load data before the experimental data was compared to the measurements. A sample graph is shown in Fig. 8.21. The spread is about 10% in the

---

[1]This code is comparable to the so-called Eppler-code [17] in Chap. 3, developed somewhat earlier and was extended to include 3D-effect from stall-delay for rotation in [32].

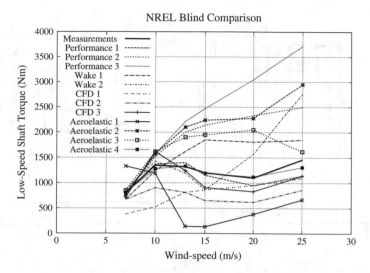

Fig. 8.21 Blind comparison of computed and measured low-speed shaft-torque for the NASA-AMES (NREL unsteady experiment phase VI) 2-bladed rotor

attached case and diverges to a factor of 7 in the stalled case. Clearly modeling of stall behavior is of utmost importance as was emphasized by Jim Tangler [56, 57]. For the first time [50] CFD gave more accurate results and details of the flow and loads that ordinary BEM and vortex methods. A detailed investigation of findings was conducted between 2003 and 2008 for IEAwind task XX, see [39] in Chap. 7.

8.6 MEXICO, New MEXICO, and MexNext

8.6.1 The 2006 Experiment

Shortly after the NASA-Ames experiment a European–Israeli project was planned in Europe's largest wind tunnel the *DNW-LLF*. See Figure 8.22. It has an open nozzle with a cross section of $9.5 \times 9.5 \, m^2$. The turbulence level was measured using pressure transducers (Kulites, used resolution about 5 kHz) giving a range of about 3‰. Much more details are described in [4, 41]. Figure 8.23 gives an impression of the setup.

The turbine has a diameter of 4.5 m and is three-bladed. As it was a co-operative European project, three profiles (see Fig. 8.24) were used:

- DU91-W2-250 from $r/R_{tip} = 0.11$ to 0.36
- RISØ A1-21 from $r/R_{tip} = 0.45$ to 0.56
- NACA64-418 from $r/R_{tip} = 0.65$ to 0.91.

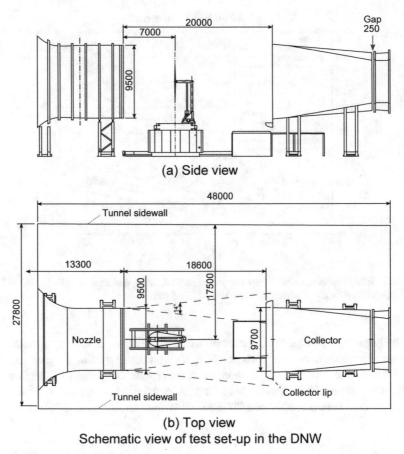

(a) Side view

(b) Top view
Schematic view of test set-up in the DNW

Fig. 8.22 Outline of the DNW-LLF wind tunnel

At about 5% chord a zig-zag tape was glued on the surface to avoid laminar separation bubbles. The chord-based Reynolds number again was in the order of 500 k for a rotational speed of 425 RPM. Unlike the NASA-Ames experiment the velocity field was also measured by *particle image velocimetry (PIV)*. This was especially useful when comparing with CFD models, see Sect. 8.6.3 and Fig. 8.25. In the upper right corner, a tip vortex seemed to emerge.

8.6.2   The 2014 Experiment

During the first two phases of MexNext (IEAwind task 29) it became clear, that agreement between CFD and experiment and accuracy (in terms of accuracy) was

Fig. 8.23 Outline of the Mexico experiment in December 2006. With permission of ECN, The Netherlands

Fig. 8.24 Lift data of used profiles

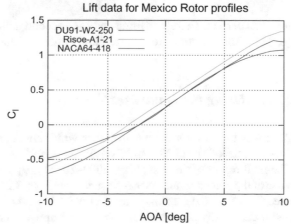

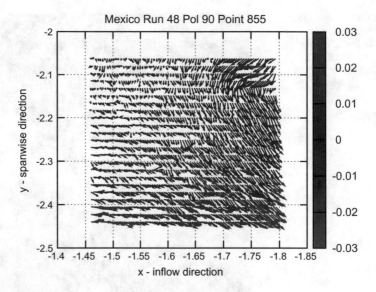

Fig. 8.25 Sample PIV data: axial traverse, pitch = 5.3°, and inflow velocity of 24 m/s, in the upper right corner a tip vortex seems to emerge

Table 8.1 Comparison of arbitrarily chosen values. Differences are in the order of 4%. See [18] for more information

Name	10 m/s		15 m/s	
	Thrust/N	Torque/Nm	Thrust/N	Torque/Nm
Mexico	854	61	1517	285
New Mexico	974	68	1163	317

less than perfect. Therefore, fortunately and second tunnel entry was possible in 2014 [6, 18]. Table 8.1 compares four arbitraily chosen values.

8.6.3   Using the Data: MexNext

As it was the case for the NASA-Ames wind tunnel experiment [39] in Chap. 7 an IEAwind task (20 and 29 respectively) was established to analyze the data in detail. The final report of the first phase (2009–2011) [29] in Chap. 5, gives an impressive view of the gained result. As may be seen from Figs. 8.26 and 8.27, much more groups used CFD than 10 years before for Annex 20. Unfortunately the spread is rather large and it can hardly be concluded that a firm improvement has been reached. A more comprehensive analysis [39] in Chap. 7 shows that most of the theoretical (BEM-like as well as CFD) models over predict the loads when compared to the measurement. Therefore, the uncertainty (systematic as well as statistical) of the experimental data

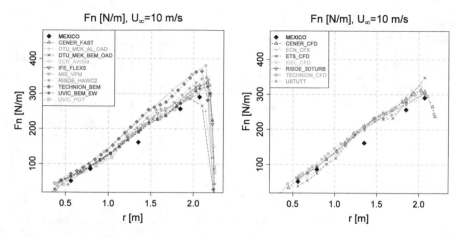

Fig. 8.26 Sample comparison (at 10 m/s inflow) of lifting line (left) and CFD (right) models in comparison to the experimental data—normal forces

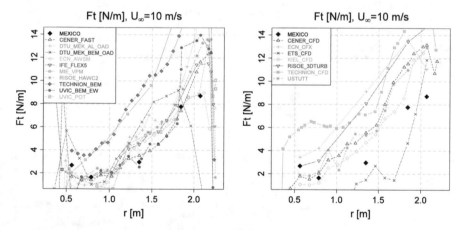

Fig. 8.27 Same as in Fig. 8.26 but for tangential forces

should be discussed. Due to the fact of the special arrangement of the inflow (nozzle) and outflow (collector), it seems difficult to fit the exact prescribed inflow conditions (see Fig. 8.28). Therefore, several authors have investigated the whole wind tunnel including the rotor modeled as an actuator disk [30, 47] and [20] in Chap. 7.

Figures 8.28 and 8.29 show some sample results. It is clearly seen that the differences for the low speed case (high tip-speed ratio λ) are most pronounced.

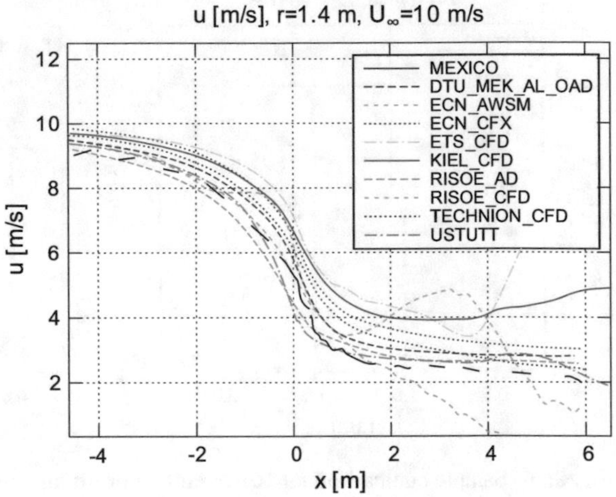

Fig. 8.28 Sample CFD calculations of various authors of the velocity field behind the Mexico-Rotor, $u_{in} = 10$ m/s, $\lambda = 10$. Data from [29] in Chap. 5

Fig. 8.29 CFD calculation of the velocity field behind the Mexico-Rotor comparing open flow and wind tunnel from [20] in Chap. 7. Same data as if Fig. 8.28

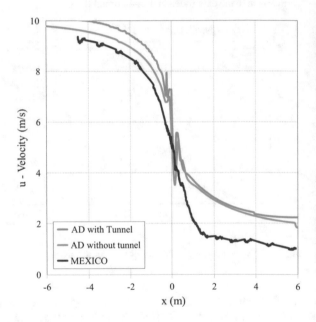

8.6.4 Wind Turbines in Yaw

A subject of continuous interest is the investigation of so-called *yawed* flow. Here wind direction and axis of rotation do not coincide. It has to be noted that yaw-control plays an important part in the total control strategy of a wind turbine. A very frequently asked question is how fast the control system should compensate for such mis-alignments. It is clear from the very beginning that because power roughly varies with yaw-angle as

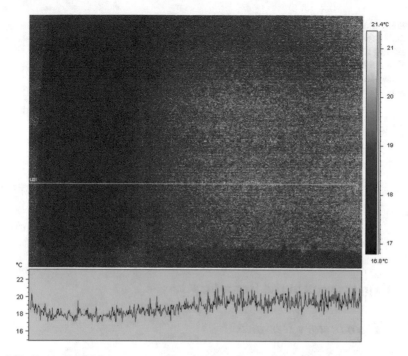

Fig. 8.30 Example of thermogrammetry in a wind tunnel experiment for a 2D airfoil. Flow from left to right. Dark areas indicate laminar parts and lighter colors turbulent flow. From [53]

$$c_{P,yawed} = c_{P0} \cdot cos(\gamma)^{n} \tag{8.1}$$

with some exponent $n \in \{2, 3\}$ [8] in Chap. 5 or even $n \in \{0 \cdots 6\}$ [5] more power could be extracted if the alignment is very strict. Unfortunately the price to pay is in form of much higher loads introduced into the azimuthal drive due to gyroscopic moments, which seem to be so high [35] that a reasonable compromise consists in very slow yaw-control only.

In addition each revolution of the rotor now produces cyclic inflow conditions. For this reason Haans, Schepers, and many others [5, 13, 35, 36, 40, 42, 44] and more recently the MexNext consortium [29] in Chap. 5, discussed yawed conditions to improve engineering guidelines [38].

In this case mesh generation is much more elaborate. Necessarily transient calculations make the turnaround time approximately one order of magnitude higher when compared to the stationary non-yawed case. A careful analysis by Schepers [42] supported by equal careful analysis of measurements on 2 MW state-of-the-art turbine indicates that tip-speed ratio is influencing the exponent γ in Eq. 8.1 strongly up to a TSR, where c_P is maximum.

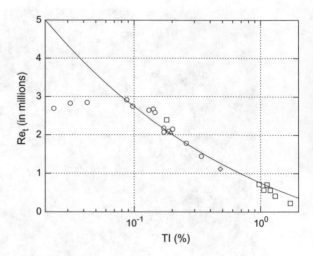

Fig. 8.31 Mack's empirical correlation for the dependency of transition-location to free-stream inflow turbulence. Data from [19]

8.7 Experiments in the Boundary Layer

8.7.1 Introductory Remarks

As we have seen in Chap. 5 the high ($c_P \approx 0.5$) values are only possible with a high lift-to-drag ratio (>100). Using standard analyzing tools like XFoil or RANS-CFD it becomes clear that both laminar (low shear stress) and turbulent (high shear stress) areas on the wing's surface must be present. Wind tunnel measurements (see Fig. 8.30) prove this, including the dependence on the turbulent inflow condition. From the computational side we have to note that most simplified transition models rely on the use of correlation [19] ($N \sim Ti$) for determining the location of a fully developed turbulent state of the boundary layer. Figure 8.31 shows measurements of transition locations on flat plates (without pressure gradients). These findings may be summarized in Eq. 8.3:

$$N = -8.43 - 2.4 \cdot ln(Ti) , \tag{8.2}$$
$$0.1 < Ti < 2\% . \tag{8.3}$$

It shows that the amplification factor N^2 varies from $N = 9$ (at 0.1% TI) up to $N = 1$ (at 2%) turbulence intensity. This lowest possible N of 1 means that the distance between first instability of laminar flow and fully developed flow (compare with Fig. 3.20) is zero. It has to be noted that this scenario only refers to the so-called *Tollmien–Schlichting* process where special wave packets occur and may be amplified. In a higher turbulent environment, this transition to turbulence may occur via *bypasses* [54].

---

[2]Roughly speaking this is the ratio where a small disturbance has grown to a relevant size.

Fig. 8.32 Example of a thermographic photograph of a rotating blade of a commercial wind turbine. Light colors indicate laminar flow and dark ones turbulent flow. Used by Permission of Deutsche WindGuard Engineering GmbH, Bremerhaven, Germany

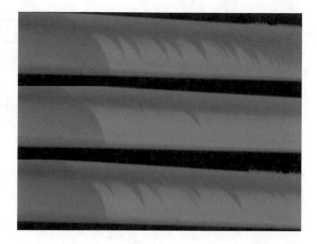

Turbulence intensities used in wind energy are mostly determined for (fatigue) load estimation and are derived by averaging over a time period of 600 s with sampling frequency of a few Hertz. They span a much wider range from 5% at minimum up to about 3% in so-called complex terrain. Clearly there should be a low-frequency cutoff in the whole energy spectrum which is aerodynamically active. Empirically this value is guessed to be on the order of 10 Hz.[3] A comparison with Fig. 3.27 shows that only a very small fraction of the total energy content is located in this frequency range. Reference [27] investigated the combined receptivity and tran.

8.7.2 Laminar Parts on the Wind Turbine Blades

In [10] the thermographic technique used in wind tunnel environments was extended for use on operating wind turbines, see Fig. 8.32. Meanwhile [31, 35] in Chap. 7 the capabilities and accuracy of thermography were investigated in much more detail. Figures 8.33 and 8.34 show for the same operational state of a Multi-MW turbine thermographic as well as microphone date on the suction side of a wind turbine blade. The first one detects transition at $x_{tr}^{Th}/c = 0.255 \pm 0.003$ and the second method $x_{tr}^{Mi}/c = 0.26 \pm 0.01$. The larger inaccuracy is mainly due to the spreading of the microphones within the array.

8.7.3 The DAN-AERO MW Experiments

A very large field experiment was undertaken during the DAN-Aero MW Experiment on state-of-the-art commercial turbines (NecMicon 2MW and Siemens SWT-2.3 both

---

[3]In [15, 16] even a much lager range of $\nu < 500$ Hz is excluded.

Fig. 8.33 Thermography

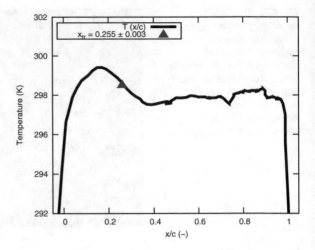

Fig. 8.34 Microphone data

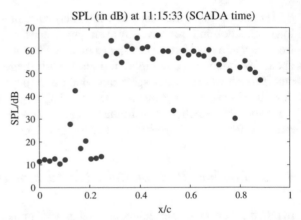

equipped with LM38.8 blades) [21, 22]. Among others, investigation of boundary layer transition was possible by use of a dense distribution of 60 microphones around a test cross section. Figure 8.35 gives an example. It has to be noted that comparable measurements on much smaller rotors (D = 25 m) have been undertaken, but were not published [15, 16]. The sudden increase from $\nu > 500\,Hz$ between POS. 1.7% and POS 2.3% may indicate transition to a turbulent flow state. Recently [24, 26] Madsen re-examined the DAN-AEro data with refined data-processing methods. Confirming most of the conclusion concerning the transition mechanism: Bypass seems more likely than Tollmien–Schlichting.

8.7.4   Hot Film Measurements

Similar experiments on a much smaller wind turbine than used in Sect. 8.7.3 [45] were performed on an ENERCON E-30 wing. For investigating the boundary layer

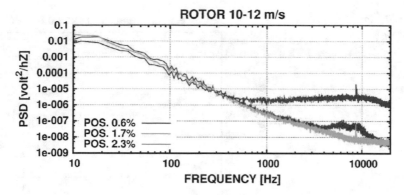

Fig. 8.35 Microphone measurements on a 38.8 m blade, data from [22, 23], used with permission of DTU wind energy

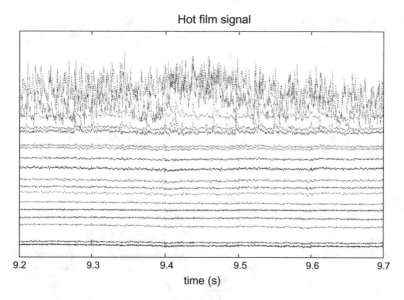

Fig. 8.36 Sample time series of a hot film located on a wind turbine blade from x/c = 0.25 (bottom) to x/c = 0.35 (top) for RN = 1 M, TI = 3%. From [45]

state a technique comparable to airplane wings was used. Those *aerodynamic gloves* experiments are common there [28, 29, 43]. Figure 8.36 shows high-frequency time series of a *hot film*. The active area was much smaller than in the experiment described in Sect. 8.7.3 and reaches from $x/c = 0.26$ (bottom) to $x/c = 0.35$ (top) only. The similarity with Fig. 3.22 (Schubauer's experiments on a flat plate) seems striking on a first glance but genuine *Tollmien–Schlichting Waves* are lacking. Figures 8.37, 8.38, and 8.39 show corresponding spectra.

Fig. 8.37 Energy spectra for
RN = 0.2 M [45]

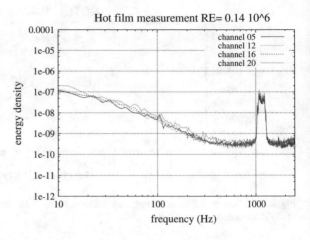

Fig. 8.38 Re = 1.1 M [45]

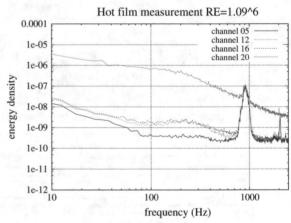

Fig. 8.39 Re = 2.8 M [45]

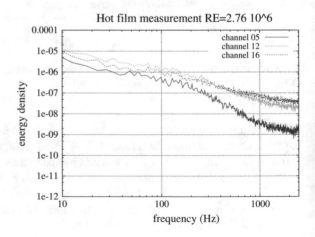

Table 8.2 Measured pressure coefficients for DU-W-300-mod

x/c	cp-0	cp-8
0.950000	0.070000	0.040000
0.900000	0.000000	−0.020000
0.850000	−0.070000	−0.110000
0.800000	−0.150000	−0.240000
0.750000	−0.220000	−0.340000
0.700000	−0.300000	−0.450000
0.650000	−0.380000	−0.560000
0.600000	−0.470000	−0.700000
0.550000	−0.550000	−0.830000
0.490000	−0.650000	−1.010000
0.440000	−0.810000	−1.170000
0.390000	−0.920000	−1.380000
0.340000	−1.020000	−1.540000
0.290000	−1.140000	−2.000000
0.240000	−1.210000	−2.140000
0.210000	−1.190000	−2.230000
0.180000	−1.140000	−2.270000
0.160000	−1.090000	−2.300000
0.140000	−1.030000	−2.300000
0.130000	−0.930000	−2.270000
0.110000	−0.860000	−2.280000
0.100000	−0.800000	−2.290000
0.090000	−0.700000	−2.260000
0.080000	−0.640000	−2.250000
0.070000	−0.570000	−2.230000
0.060000	−0.480000	−2.210000
0.050000	−0.400000	−2.190000
0.040000	−0.320000	−2.160000
0.040000	−0.210000	−2.140000
0.030000	−0.110000	−2.120000
0.020000	−0.040000	−2.110000
0.020000	0.070000	−2.080000
0.010000	0.180000	−2.030000
0.010000	0.320000	−1.970000
0.010000	0.440000	−1.880000
0.000000	0.570000	−1.780000
0.000000	0.740000	−1.540000

(continued)

Table 8.2 (continued)

x/c	cp-0	cp-8
0.000000	0.870000	−1.190000
0.000000	0.980000	−0.660000
0.000000	1.000000	−0.350000
0.000000	0.990000	0.280000
0.010000	0.810000	0.810000
0.020000	0.600000	0.980000
0.030000	0.370000	0.970000
0.050000	0.130000	0.880000
0.070000	−0.140000	0.710000
0.100000	−0.460000	0.450000
0.150000	−0.840000	0.090000
0.240000	−1.520000	−0.550000
0.310000	−1.450000	−0.700000
0.380000	−1.090000	−0.500000
0.450000	−0.570000	−0.180000
0.520000	−0.260000	−0.020000
0.580000	−0.060000	0.120000
0.650000	0.130000	0.250000
0.720000	0.250000	0.330000
0.860000	0.370000	0.430000
0.930000	0.410000	0.460000
1.000000	0.130000	0.060000
0.000000	0.000000	0.000000

8.8 Summary of Experiments in Wind Turbine Aerodynamics

It has been shown that experiments even only on small parts of a wind turbine are extensive and complicated. Also it has been stated many times that—compared to aerospace industry—wind turbine research is comparatively young. Therefore, it may be easy to predict that new experiments both in wind tunnels and in the field will conducted in the near future and will bring much progress in scientific and practical engineering.

8.9 Problems

Problem 8.1 Use the c_p data given in Table 8.2 to integrate for lift.

Problem 8.2 Download data for the NTk Wind turbine from **wind data.com** to estimate a power curve.

Problem 8.3 Use measured data of Problem 8.1 to prepare plots of lift-to drag ratio.

References

1. Abbot IH, von Doenhoff AE (1958) Theory of Wing Sections. Dover Publication Inc, New York
2. Ahmad MM (2014) CFD investigations of the flow over FLAT Back Airfoils using OpenFOAM and different turbulence models. MSc thesis, UAS Kiel and FhG IWES, Kiel and Oldenburg, Germany
3. NN (2000) Basic machine parameters. Paper circulated during NASA Ames blind comparison panel, NREL, Golden, USA
4. Boorsma K, Schepers JG (2011) Description of experimental setup - MEXICO measurements, ECN-X-11-120, Confidential, ECN, Petten, The Netherlands
5. Boorsma K (2012) Power and loads for wind turbines in yawed conditions, ECN-E-12-047, ECN, Petten, The Netherlands
6. Boorsma K, Schepers JG (2018) Description of experimental setup, New Mexico Experiment, version 3 ECN-X-15-093, Petten, The Netherlands
7. Björck A, Ronsten G, Montgomerie B (1995) Aerodynamic section characteristics of a rotating and non-rotation 2.375 m wind turbine blade, FFA TN 1995-03, Bromma, Sweden
8. Butterfield CP, Musial WP, Scott GN, Simms DA (1992) NREL combined experimental final report - Phase II, NREL/TP-442-4807, Golden, CO, USA
9. Dexin H, Thor S-E (1993) The execution of wind energy projects 1986–1992, FFA TN 1993–19, Bromma, Sweden
10. Dollinger Chr, Balaresque N (2013) Messverfahren zur akustisch-aerodynamischen Optierung von Rotorblättern im Winkanal, priv. comm. (in German)
11. Elsamprojker A/S (1992) The Tjaæreborg wind turbine, Final Report, CEC, DG XII, contract EN3W.0048.DK, Fredericia, Denmark
12. Freudenreich K, Kaiser K, Schaffarczyk AP, Winkler H, Stahl B (2004) Reynolds number and roughness effects on thick airfoils for wind turbines. Wind Eng 28(5):529–546
13. Haans W (2011) Wind Turbine Aerodynamics in Yaw. PhD thesis, TU Delft, Delft, The Netherlands
14. International Electro-technical Commission (2013) IEC 61400–12-2, wind turbines - part 12–2: power performance of electricity producing wind turbines based on nacelle anemometry, Switzerland, Geneva
15. van Groenwoud GJH, Boermans LMM, van Ingen JL (1983) Onderzoek naar de omslag laminair-turbulent van de grenslaag op de rotorbladen vand de 25 m HAT windturbine, Rapport LR-390. Techische Hogeschool Delft, Delft, The Netherlands
16. van Ingen JL, Schepers JG (2012) Prediction of boundary layer transition on wind turbine blades using e^{N}-method and a comparison with measurements, private communication, G Schepers
17. Lissaman PBS (1983) Low-Reynolds-number airfoils. Ann Rev Fluid Mech 15:223–239
18. Phisipsen I, Heinrich S, Pengel K, Holthusen H (2015) Test report for measurements on the Mexico wind turbine model in DNW-LLF LLF-2014-19, Marnesse, The Netherlands
19. Mack LM (1977) Transition and laminar instability, 77–15. JPL Publication, Pasadena
20. Madsen J, Lenz K, Dynampally P, Sudhakar P (2009) Investigation of grid resolution requirements for Detached Eddy simulation of flow around thick airfoil sections. In: Proceedings of EWEC 2009, Marseille, France

21. Madsen HA et al (2009) The DAN-AERO MW experiment final report, Risø-R-1726(EN), Roskilde, Denmark
22. Madsen HA et al (2010) The DAN-AERO MW experiment, AIAA-2010-645, Orlando, FL, USA
23. Madsen HA, Bak C (2012) The DAN-AERO MW experiment, IEAwind Annex 29 (MeNext) annual meeting, Golden, CO, USA
24. Aa H, (2019) Madsen Transition, characteristics measured on a 2 MW 80m diameter wind turbine rotor in comparison with transition data from wind tunnel measurement, AIAA-2019-0801, AIAA Scitech, et al (2019) Formum. San Diego, CA, USA, p 2019
25. Özlem CY, Pires O, Munduate X, Sørensen N, Reichstein T, Schaffarczyk AP, Diakakis K, Papadakis G, Daniele E, Schwarz M, Lutz T, Prieto R (2017) Summary of the blind test campaign to predict high reynolds number performance of DU00-W-210 airfoil. AIAA 2017–0915:915
26. Özçakmak ÖS, Sœrensen NN, Madsen HA, Sœrensen JN (2019) Laminar-turbulent transition detection on airfoils by high-frequency microphone measurements. WIND ENERGY 22:10. https://doi.org/10.1002/we2361
27. Ohno D, Romblad J, Rist U (2020) Laminar to turbulent transition at unsteady inflow coditions: numerical simulations with small scale free-stream turbulence, In: Dillmann A et al (eds) DGLR 2018, NNFM 142, pp 214–225
28. Peltzer I et al (2009) In flight experiments for delaying laminar-turbulent transition on a laminar wing glove. Proc. IMechE 223:619–626
29. Reeh AD, Weissmüller M, Tropea C (2013) Free-flight investigations of transition under turbulent conditions on a Laminar wing glove, AIAA-2013-0994, Grapevine, TX, USA
30. Réthoré P-E et al (2011) MEXICO wind tunnel and wind turbine modeled in CFD, AIAA-3373, Orlando, FL, USA
31. Ronsten G (1992) Static pressure measurements on a rotating and a non-rotating 2.375 m wind turbine blade. Comparison with 2D calculations. J Wind Eng Ind Aerodyn 39:105–118
32. van Rooij RPJOM (1996) Modifications of the boundary layer calculation in RFOIL for improved airfoil stall prediction, report IW-96087R, TU Delft, Delft, The Netherlands
33. van Rooij RPJOM (2007) Open air experiments on rotors. In: Brouckert J-F (ed) Wind turbine aerodynamics: a state-of-the-art, Lecture series 2007–05, von Karman institute for fluid dynamics. Rhode Saint Genese, Belgium
34. Schaffarczyk AP (2008) Numerische Polare eines 46% dicken aerodynamischen Profils, Bericht des Labors für Numerische Mechanik, 58, Kiel Germany (in German, confidential)
35. Schaffarczyk AP (2011) Expertise zum Einsatz eines Lasermesssystems zur Verbesserung des Energieertrages und Reduzierung der Lasten mittels genauerer Windnachführung einer Windenergieanlage (Use of a Laser system for increased energy yield and load reduction by improved yaw control), report No. 83, Kiel, Germany (in German, confidential)
36. Schepers JG, Snel H (1995) Dynamic Inflow: Yawed Conditions and partial span pitch control, ECN-C-95-056. Petten, The Netherlands
37. Schepers JG et al (1997) Final Report of IEA ANNEX XIV, Field Rotor Aerodynamics, ECN-C-97-027, Petten, The Netherlands
38. Schepers JG (1999) An engineering model for yawed conditions, developed on the basis of wind tunnel measurements. AiAA-paper 1999–0039:164–174
39. Schepers JG et al (2002) Final Report of IEA ANNEX XVIII, 'Enhanced Field Rotor Aerodynamics Database, ECN-C-02-016, Petten, The Netherlands
40. Schepers JG (2004) ANNEXLYSE: Validation of yaw models, on basis of detailed aerodynamic measurements on wind turbine blades, ECN-C-04-097, ECN, Petten, The Netherlands
41. Schepers JG, Snel H (2007) Model experiment in controlled conditions - Final Report, ECN-E-07-042, Petten, The Netherlands
42. Schepers JG (2012) Engineering models in wind energy aerodynamics. PhD thesis, TU Delft, Delft, The Netherlands
43. Seitz A (2007) Freiflug-Experimente zum Übergang laminar-turbulent in einer Tragflügelgrenzschicht, DLR-FB-2007-01, Braunschweig, Germany (in German)

44. Snel H, Schepers JG (1995) Joint investigation of Dynamic Inflow Effects and implementation of an engineering method, ECN-C-94-056, Petten, The Netherlands
45. Schwab D, Ingwersen S, Schaffarczyk AP, Breuer M (2012) Pressure and hot film measurements on a wind turbine blade operating in the atmosphere. In: Proceedings of the science of making torque from wind, Oldenburg, Germany
46. Shen WZ, Hansen MOL, Sœrensen JN (2009) Determination of the angle of attack on rotor blades. Wind Energy 12:91–98
47. Shen WZ, Zhu WJ, Sørensen JN (2012) Actuator line/Navier-Stokes computations for the MEXICO rotor: comparison with detailed measurement. Wind Energy 15:151–169
48. Simms DA, Hand MM, Fingersh LJ, Jager DW (1999) Unsteady aerodynamics experiment phases II-IV, test configurations and available data campaigns, NREL/TP-500-25950. Golden, CO, USA
49. Simms D, Schreck S, Hand M, Fingersh LJ (2001) NREL unsteady aerodynamics experiment in the NASA-Ames wind tunnel: a comparison of predictions to measurements, NREL/TP-500-29494. Golden, CO, USA
50. Sœrensen NN, Michelsen JA, Schreck S (2002) Navier-Stokes prediction of the NREL phase VI rotor in the NASA Ames 80 ft × 120 ft wind tunnel. Wind Energy 5:151–169
51. Somers D (1997) Design and experimental results for the S809 airfoil, NREL/SR-440-6918, Golden, CO, USA
52. Stahl B, Zhai J (2003) Experimentelle Untersuchung an einem 2D-Windkraftprofil im DNW-Kryo Kanal, DNW-GUK-2003 C 02, Köln, Germany (in German)
53. Stahl B, Zhai J (2004) Experimentelle Untersuchung an einem 2D-Windkraftprofil bei hohen Reynoldszahlen im DNW-Kryo Kanal, DNW-GUK-2004 C 01, Köln, Germany (in German)
54. Suder KL, OBrian JE, Roschko E, (1988) Experimental study of bypass transition in a boundary layer, NASA, Technical Memorandum 100913, Cleveland, Ohio, USA
55. Timmer WA, Schaffarczyk AP (2004) The effect of roughness at high Reynolds numbers on the performance of airfoil DU9 97-W-300Mod. Wind Energy 7(4):295–307
56. Tangler JL (2004) The Nebulous art of using wind-tunnel airfoil data for predicting rotor performance, NREL/CP-500-31243, Golden Co, USA
57. Tangler JL, Kocurek JD (2004) Wind turbine post-stall airfoil performance characteristics guidelines for blade-element momentum methods, NREL/CP-500-36900, Golden Co, USA
58. Wolf M, Jeromin A, Schaffarczyk AP (2010) Numerical prediction of airfoil aerodynamics for thick profiles applied to wind turbine blade roots. In: Proceedings of the DEWEK 2010, Bremen, Germany
59. Zell PT (1993) Performance and test section flow characteristics oft he national full-scale aerodynamics complex 80- by 120-foot wind tunnel, NASA Technical Memorandum, 103920, Moffett Field, CA, USA

Chapter 9
Impact of Aerodynamics on Blade Design

> *So wie die Sache steht, ist das Beste, auf das zu hoffen ist, ein Geschlecht erfinderischer Zwerge, das für alles zu mieten ist (B. Brecht, Life of Galileo, 1941), [6]. (As things are, the best that can be hoped for is a generation of inventive dwarfs who can be hired for any purpose.)*

9.1 The Task of Blade Design

Now having introduced all knowledge from fluid mechanics, it is high time to try to give an overview on what is really used in practical wind turbine blade design. Referring to Chap. 10 and esp. Fig. 10.1 we see that with the development of huge offshore wind turbines up to 170 m rotor diameter, a period of exponential growth has started again after some years of dormancy. A European Project *UpWind* [27] (compare to [17]) forecasts feasible blades for wind turbines up to 20 MW with blade lengths up to 120 m, while currently (2014) turbines reaching rotor diameters of 170 m and rated power of 8 MW have been constructed. Design of a real commercial wind turbine blade is a complicated process and involves (at least)

- aerodynamics,
- structural design, and
- manufacturing.

Aerodynamics therefore only touches the outer skin of the blade giving an idea how the shape should be. Manufacturing relies very much on the structural design, because this more or less defines the amount of material used. On the other hand the amount of material used determines mass and this is related by engineering rules to costs and prices [28]. To give an order of magnitude, €10 per kg have not be exceeded to be competitive on the market place. To see how masses evolve or *scale*, we may have a look on Figs. 9.1, 10.2 and 2.6 of Chap. 10.

A double logarithmic plot reveals the scaling exponent n from $m \sim D^n$. It must clearly lie between 2 and 3: in the first dimension, the length of the blade is determined by the gross design of the rotor. Then the chord c follows from $\Gamma \sim c \cdot c_L$

© Springer Nature Switzerland AG 2020
A. P. Schaffarczyk, *Introduction to Wind Turbine Aerodynamics*,
Green Energy and Technology, https://doi.org/10.1007/978-3-030-41028-5_9

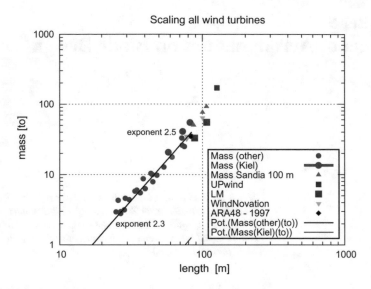

Fig. 9.1 Scaling of wind turbine blade masses in metric tons

and yields—by an appropriate airfoil choice—(see Sect. 9.2) the second dimension. The remaining third dimension then is left to the structural engineer. Scaling only length and chord without augmenting the internal structure then results in $n = 2$, whereas simply increasing **all three** dimensions results in $n = 3$ [16]. From Fig. 9.1 and Chap. 5, [16] we may see that smaller ($R_{tip} < 80\ m$) blades have had $n \leq 2.4$ where larger blades seem to need so much more material resulting in $n > 3$ [17, 28]. The reason for this is rather simple: structural design according to the international standard of [26] designate loads that the structure has to withstand for design lifetime of 20 years, for example. We will come back to these items in Sect. 9.4.

9.2 Airfoils for Wind Turbine Blades

Choosing the right airfoil from an *airfoil catalogue*, see Chap. 8, [1] or [1, 24], comes next after having chosen blade length and having identified lift c_L and chord c. Numerous profiles exist but only some measured aerodynamic lift data have even been published. As a general introduction the reader may consult [22]. To minimize losses, lift-to-drag ratio has to be as high as possible, as we saw in Chap. 5. Nevertheless the highest forces are associated with c_L^{max} so this has be be controlled during the design phase and confirmed by measurements.

The inverse problem, designing the shape of a profile when the (inviscid) pressure distribution around the profile is given can be performed by *conformal mapping* (Chap. 3, [17]) and was used by Tangler and Somers (NREL) [30, 31] for designing

Fig. 9.2 Flat-back airfoil
from University of Stuttgart
[1]

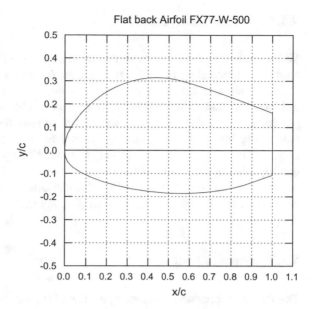

the Sxxx family of airfoils. One of them (S809) has been used for the NREL NASA-Ames experiment as already described in Sect. 8.5.

Somewhat earlier (in the late 70s) during the *GROWIAN* program [32] at the University of Stuttgart, Wortmann(FX), and Althaus(AH) [1] profiles were investigated.

When XFoil Chap. 3, [14] become popular in the late 90s at the Aeronautical Research Association in Sweden (FFA, [3, 4, 13]) and at the Technical University Delft in the Netherlands [33, 34], special profiles were investigated carefully. Risø (now DTU Wind energy) [12] did the same. Only very limited information is freely available as to how much these genuine wind turbine blade profiles are used in commercial blades.

Sometimes very slim blades [7, 11] are considered for blade design in combination with *high-lift airfoils*.

Some thick ($\geq$30% thickness, see Figs. 8.4, 8.5, and 8.6) airfoils for root sections have already been presented in Chap. 8. There seems to be still room for further improvement [15].

Very recently so-called *flat-back airfoils*, see Fig. 9.2 [2] have been re-discovered again.

9.3   Aerodynamic Devices

Aerodynamic devices are small devices which may easily be added or removed from the wing. Among them two play a noteworthy role:

- Vortex generators and
- Gurney flaps.

9.3.1   Vortex Generators

Vortex generators (VGs) are common from aerospace industry. Their use and the-oretical understanding is quite extensive, such that a meaningful use is possible. Figure 9.3 shows a typical overview of an arrangement used in medium-sized wind turbines. The triangle extensions are $40 \times 10 \, mm^2$. They are used mainly to introduce (additional) lift without regard to the additional drag they produce. This is clearly seen in Fig. 9.4: c_L^{max} is increases by about 20% but the drag as well, so that L2D decreases from about 80 (clean blade)[1] to about 45 only. VGs have been used from the beginning of modern wind turbine applications on blades, mainly to prevent the part close to the hub from early separation. Details of the flow pattern behind VGs have been investigated recently by Velte [35] and others.

9.3.2   Gurney Flaps

This device—apparently invented by accident [19]—consists of a small trailing edge flap, see Fig. 9.5. Because of its simplicity to increase lift, it may be used to compen-sate fouled profiles. In Fig. 9.6 results are shown where the fouling (by insects, sand or other influences) was modeled by some defined surface roughness (Karborun-dum). Figure 9.6 show the results of attempts to improve for higher lift. The impact on lift-to-drag ratio is summarized in Fig. 9.7. Only in combination with Gurney flaps, a small improvement in terms of lift-to-drag ratio seems to be possible when a severe roughness close to the nose is present.

9.4   Structural Design and Manufacturing

9.4.1   Structural Design

Wind turbines are used in most cases to produce electricity. The first part of a design therefore is to ensure the predicted power performance. Next to this—and probably much more important—is to ensure *structural integrity* during the foreseen lifetime which—in most cases—is at least 20 years. As we have seen the primary load is the thrust of a wind turbine, Eq. (3.20). Guidelines and regulations for a safe structural design have evolved gradually [8, 14, 26]. Apart from aerodynamics, knowledge from rotor dynamics [36] and structural design [9][2] is necessary. Some of the tasks are to

---

[1] This open-jet wind tunnel has a comparably high turbulence intensity of more than 1%.

[2] This textbook from the late 1980s is surely out-dated but may help to bridge the gap between engineering mechanics and black-box tools like FLEX5 or BLADED.

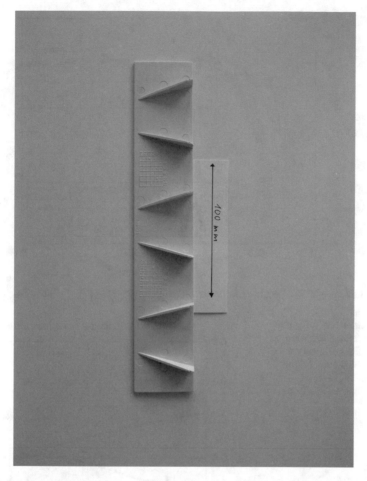

Fig. 9.3 Example of vortex generators used on wind turbine blades. Flow from left. Photo: Schaffarczyk

- Ensure safety against aeroelastic instabilities such as divergence/flutter,
- Investigate loads from extreme events expected only a few times within the estimated lifetime,
- Sum the effects of loads from rapidly changing operating conditions during normal or electricity generation.

A typical iterative process is shown in Fig. 9.8. A turbine may be called optimized if it can resist equally extreme winds and fatigue loads. The IEC standard [26] proposes three possibilities for proof of safety:

- a so-called *simplified load model*,
- simulation modeling (by aeroelastic codes), and
- load measurement.

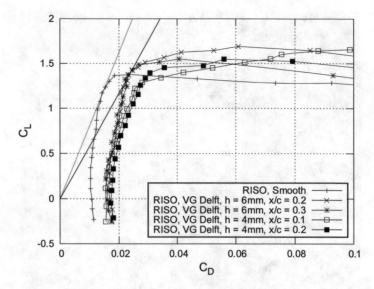

Fig. 9.4 Influence of VGs on the performance on the FFA-W3-211 [13]

Fig. 9.5 Sketch of a Gurney flap used in the 2002/2003 Kölner KK experiment on high Reynolds number flow on the DU-W-300-mod profile (Chap. 8, [52, 53])

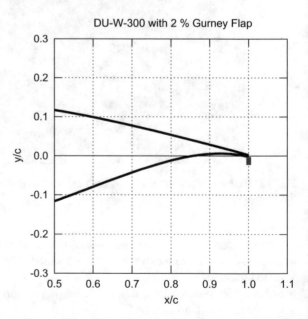

To compare the amount of work required it may be interesting to note that the first model is available as Excel spreadsheet [37], and a safety check may be finished in hours. A full aeroelastic load model takes much more time (may months), and clearly for measurements a full prototype must be available.

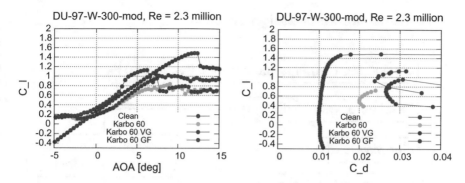

Fig. 9.6 Influence of various devices on DU97-W-300-mod measured at Kölner KK, Re = 2.3 M

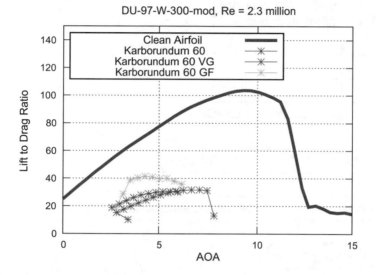

Fig. 9.7 Influence of various measures on lift-to-drag ratio. Karbo = Karborundum = severe roughness in the form of sand. VG = Vortex generators, GF = Gurney flaps

Aeroelastic modeling may be identified at different stages of sophistication:

I Aerodynamic

1. Blade-Element-Momentum Code (BEM)
2. Wake Codes
3. Full 3D CFD including turbulence modeling

II Structural Mechanics

4. Beam (1D) model
5. Shell (2D) model
6. Solid (3D) model

Blade structural design process

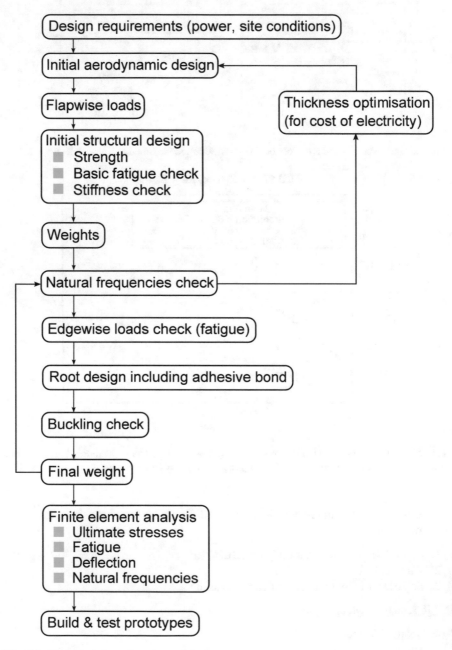

Fig. 9.8 Simplified chart of a structural design

Fig. 9.9 Drag coefficient of cylindrical shells. Measured data from Wieselsberger. Turbulent and transitional CFD simulations

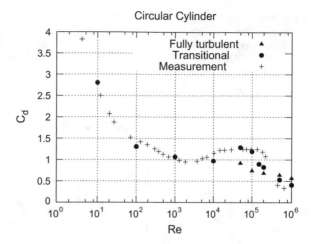

In principle $3 \cdot 3 = 9$ possibilities for coupling the various methods can occur. At the present time (1 with 4) coupling is the one most commonly used in industry. Several industrial codes are available, some of which are

- FLEX, from Stig Øye, DTU,
- BLADED, by Garrad Hassan,
- GAROS, by Arne Vollan, FEM,
- Phatas, by ECN,

and many more.

One special item of structural design in connection with ultimate loads [26] and more directly related to aerodynamics may be presented in more detail here. Figure 9.10 shows radially resolved drag coefficients which are important in standstill conditions during severe storms. It can be seen that—compared to older blades— c_D is decreased in the inner parts(closer to hub) and increased in the thinner outer parts (closer to tip). As Reynolds numbers then vary between 500 k and 5 M, predictions must be capable to model the so called *drag crisis*. Modern CFD codes with transition modeling included seem to be able to do so, see Fig. 9.9. It may be seen from Fig. 9.10 that newer blades try to reduce these loads from the inner part of the blade to the tip.

9.4.2 Manufacturing

Wind turbine blades are manufactured mostly from *glass (or carbon) fiber reinforced plastics (GFRP)* [18, 23]. A typical process is to use moulds which are separated into a lower and upper part. The price for such a mould system varies based on the number of actually manufactured blades between one and several millions Euros. Reference

Fig. 9.10 Radial resolved
drag coefficient of two wind
turbine blades: LM 19—an
old stall blade and more
modern blades

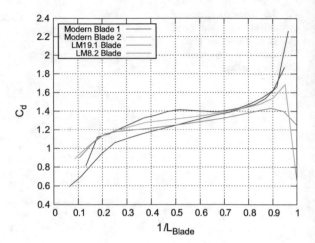

[23] gives a detailed sample calculation for a 55 m blade of about 18 metrics to mass. Material and labor costs (€45 per hour) are of almost equal share to give a total production price of about €160 T corresponding to €8.9 per kilogram. This price is much lower than typical prices for comparable parts used in commercial airplanes. Nevertheless, and due to the fact that the price per produced kWh has to be decreased further, blades have to become even cheaper. Large-scale production and automation of manufacturing are the key issues for achieving these goals. The normalized price of wind turbines, see Fig. 10.3, will be presented in more detail in Chap. 10.

9.5 Examples of Modern Blade Shapes

Figure 9.11 shows cross sections for an active stall blade from the late 1990s producing about 0.5 MW rated power. It was presented in some detail in Chap. 7. The aerodynamic experts used profiles from Chap. 8, [1] and BEM methods improved by empirical and so-called 3D corrections. This rated power was state-of-the-art back then. Approximately 20 years later, two to three MW now is state-of-the-art for onshore turbines. Figures 9.12 and 9.13 give some impression for blade shapes from German manufacturer ENERCON. As is well known, ENERCON use direct-driven (no gear box) turbines which requires larger nacelle diameters. It is therefore pertinent to design the inner part of the rotor interlocking with the nacelle. This program already started in 2002 [25] and a new type E-70 (rated power 2.0–2.3 MW) was released in 2004. Unfortunately very limited information on the aerodynamic properties is available for these kind of manufacturer-owned blade designs.

In 2014 the largest manufactured blades [10] had a length of 83 m and a mass of less than 35 metric tonnes. They belong to a 7 MW (S7.0-171) rated power offshore wind

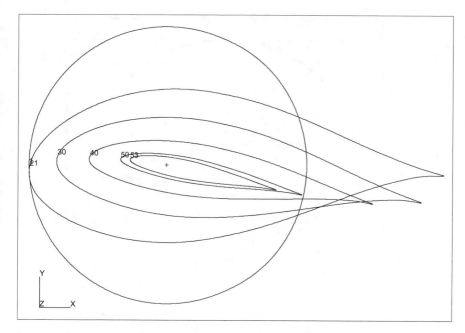

Fig. 9.11 Cross section of an active stall blade. Rated power around 0.5 MW, length 25 m

Fig. 9.12 Rated power around 2 MW, diameter 82 and 92 m, respectively, Reproduced with permission from ENERCON GmbH, Aurich, Germany

Fig. 9.13 Rated power around 3 MW, diameter 101 and 115 m, respectively, Reproduced with permission from ENERCON GmbH, Aurich, Germany

turbine from Samsung Heavy Industries, see Fig. 9.15. Partially flat-back airfoils are used, see Fig. 9.14.

By now (summer 2019) the largest manufactured blades have a length of 107 m (see Fig. 9.16) and an estimated mass between 55 and 63 metric tonnes. They belong to a 12 MW (GE Haliade-X 12 MW) rated power offshore wind turbine from GE.

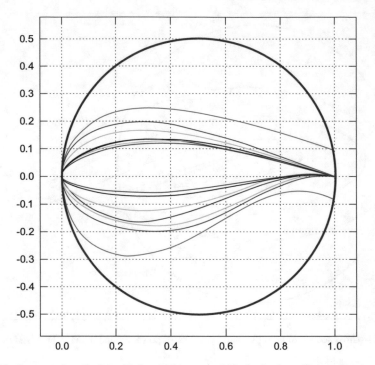

Fig. 9.14 Cross section of a large blade which uses partially flat-back profiles

There are further projects in research (Avatar, INNwind), see Chap. 10. Industry certainly will bring additional ongoing progress.

9.6 Problems

Problem 9.1 Design blades for a wind car (see Problem 5.5)[3] which is specified as follows: $B = 3$, TSR $= \lambda = 5.5$, L2D-ratio $= 80$, $R_{tip} = 0.9$ m, $c_{L,des} = 1.0$ at AOA $= \alpha_{des} = 6° = 6$ deg, $\tilde{w}/u_1 = 0.5$, and drivetrain efficiency $\eta_{DrTr} = 0.7$.

---

[3] Much of the underlying principle (low induction) has recently been re-introduced for the design of 10+ MW ordinary wind turbines (Chap. 10, [3]).

Fig. 9.15 Largest (R = 83 m) wind turbine blade in spring 2014 (1st edition). Reproduced with permission from SSP Technology A/S, Stenstrup, Denmark

Fig. 9.16 Largest (R = 107 m) wind turbine blade (summer 2019) thus far. Reproduced with permission from LM

References

1. Althaus D (1996) Niedriggeschwindigkeitsprofile: Profilentwicklungen und Polarenmessungen im Laminarwindkanal des Institutes für Aerodynamik und Gasdynamik der Universität Stuttgart. Vieweg, Braunschweig (in German)
2. Baker JP, van Dam CP, Gilbert BL (2008) Flat-back airfoil wind tunnel experiment, SAND2008-2008, Sandia National Laboratories, Albuquerque, NM, USA
3. Björk A (1990) Coordinates and calculations for the FFA-W1-xxx, FFA-W2-xxx and FFA-W3-xxx series of airfoils for horizontal axis wind turbines, FFA TN 1990–15, Stockholm, Sweden
4. Björk A (1996) A guide to data files from wind tunnel test of a FFA-W3-211 airfoil at FFA, FFAP-V-019, Stockholm, Sweden
5. Boorsma K, Machielse L, Snel H (2010) Performance analysis of a shrouded rotor for a wind powered vehicle. In: Proceedings of the TORQUE 2010, Crete, Greece
6. Brecht B (2008) Life of Galileo. Penguin Classics, London, Reprint
7. Corten G (2007) Vortex blades - proposal to decrease turbine loads by 5%, WindPower 2007, Los Angeles, CA, USA
8. Det Norske Veritas(DNV)/Riso/o (2002) Guidelines for design of wind turbines, 2nd edn. Roskilde, Denmark
9. Eggleston DM, Stoddard FS (1987) Wind turbine engineering design. van Nordstand, New York
10. Eichler K (2013) SSP technology - blade design. In: VDI-conference, rotor blades of wind turbines, Hamburg, Germany, 17–18 April 2013
11. Fuglsang P (2004) Aero-elastic blade design - slender blades with high lift airfoils compared to traditional blades. Wind turbine blade workshop, Albuquerque, NM, USA
12. Fuglsang P, Bak C (2004) Development of the Risø wind turbine airfoils. Wind Energy 7(2):145–162
13. Fuglsang P et al (1998) Wind tunnel tests of the FFA-W3-241, FFA-W3-301 and NACA 63–430 airfoils, Risø-R-1041(EN), Roskilde, Denmark
14. Germanischer Lloyd Windenergie GmbH (2010) Guidelines for the certification of wind turbines, Hamburg
15. Grasso F (2012) Design of thick airfoils for wind turbines. Wind turbine blade workshop, Albuquerque, NM, USA
16. Griffith DT, Ashwill D (2011) The Sandia 100-meter all-glass baseline wind turbine blade: SNL 100-00, SAND2011-3779, Sandia National Laboratories, Albuquerque, NM, USA
17. Hillmer B et al (2007) Aerodynamic and structural design of MultiMW wind turbine blades beyond 5 MW. J Phys Conf 75:01202
18. Jaquemotte P (2012) Fertigung von Rotorblättern, Einsatz von Carbonfasern (Manufacturing of rotor blades - use of carbon fibers), Private communication (in German)
19. Katz J (2006) Aerodynamics of race cars. Annu Rev Fluid Mech 38:27–63
20. Lehmann J, Kühn M (2009) Mit dem Wind gegen den Wind. Das Windfahrzeug InVentus Ventomobil. Physik in unserer Zeit 4:176–181
21. Lehmann J, Miller A, Capellaro M, Kühn M (2008) Aerodynamic calculation of the rotor for a wind driven vehicle. In: Proceedings of the DEWEK, Bremen, Germany
22. Lissaman PBS (2009) Wind turbine airfoils and rotor wake. In: Spera DA (ed) Wind turbine technology, 2nd edn. ASME Press, New York
23. Ludwig N (2013) Automated processes and cost reductions in rotor blade manufacturing. In: VDI-conference, rotor blades of wind turbines, Hamburg, Germany, 17–18 April 2013
24. Miley SJ, A catalog of low Reynolds number airfoil data for wind turbine applications, RFP-3387 UC-60, Golden, CO, USA
25. NN (2004) New rotor-blades - innovative feature, Private communication
26. NN (2007) IEC publication 61400-1, 3rd edn. International Electro-technical Commission, Geneva

27. NN (2011) UpWind - design limits and solutions for very large wind turbines, EWEA, Brussels, Belgium
28. Sieros G et al (2012) Upscaling wind turbines: theoretical and practical aspects and their impact on the cost of energy. Wind Energy 15(1):3–17
29. Sørensen JN. Aero-mekanisk model for vindmølledrevet køretøj, Unpublished report, Copenhagen, Denmark (in Danish)
30. Tangler J, Somers DM (1995) NREL airfoil families for HAW turbine dedicated airfoils. In: Proceedings of the AWEA
31. Tangler J, Smith B, Jager D (1992) SERI advanced wind turbine blades, NREL/TP-257-4492, Golden, CO, USA
32. Thiele HM (1983) GROWIAN-Rotorblätter: Fertigungsentwicklung, Bau und Test (GROWIAN's rotorblades: development, construction and testing), BMFT-FB-T 83–11, Bonn, Germany (in German)
33. Timmer WA (2007) Wind turbine airfoil design and testing. In: Brouckert J-F (ed) Wind turbine aerodynamics: a state-of-the-art. Lecture series 2007–05. von Karman Institute for Fluid Dynamics, Rhode Saint Genese
34. Timmer WA, van Rooij RPJOM (2003) Summary of the Delft University wind turbine dedicated airfoils. J Sol Energy Eng 125(4):488–496
35. Velte CM (2009) Characterization of vortex generator induced flow. PhD thesis, Technical University of Denmark, Lyngby, Denmark
36. Vollan A, Komzsik L (2012) Computational techniques of rotor dynamics with the finite element method. CRC Press, Boca Raton
37. Wood D (2011) Small wind turbines. Springer, London

Chapter 10
Concluding Remarks on Further Developments

10.1 State of the Art

In this book, we have tried to give an overview of aerodynamics of wind turbines and its application to blade design. During the last few decades rotor size (see Fig. 10.1) increased more than one order of magnitude (exponentially). After a five-year period of stagnation, starting in 2010 there seems to be exponential rotor size growth occurring again culminating so far with design and construction of 2220 m diameter 12 MW offshore wind turbine.

Nevertheless design methods developed in the first third of the last century (BEM) are still state of the art for industry. Progress to more sophisticated fluid mechanics (CFD) is slow partly due to the high amount of computational effort and partly due to the limited gain in accuracy. It has to be noted that blade masses have shown to be much smaller than originally estimated see [7] and Fig. 9.1 from Chap. 9. Advanced control plays a key role but any details are out of the scope of this textbook.

10.2 Upscaling

In Chap. 9, Fig. 9.1 we have already seen a general trend for upscaling when including all sizes of wind turbines. In Fig. 10.2 only *big* (larger than state of the art, larger than 2–3 MW rated power, $R > 60$ m) turbines are shown. Early attempts of UAS Kiel ([17] in Chap. 9) for a feasible 10 MW blade design resulted in a *scaling exponent* $m \sim R^n$ of $n \approx 2.5$. Sandia ([16] in Chap. 9) simply **used** $n = 3$ for their approach. Very recent design ([12] in Chap. 9) and theoretical investigations [1, 3], see Fig. 10.2 show that a even smaller ($n \approx 2.2$) should be possible. They based their estimates on glass-fiber blade data for lengths between 40 and 80 m.

Since 2006 many projects were started for investing in turbines up to 20 MW rated power. The first of them called *UPwind* lasted from March 2006 to February

© Springer Nature Switzerland AG 2020
A. P. Schaffarczyk, *Introduction to Wind Turbine Aerodynamics*,
Green Energy and Technology, https://doi.org/10.1007/978-3-030-41028-5_10

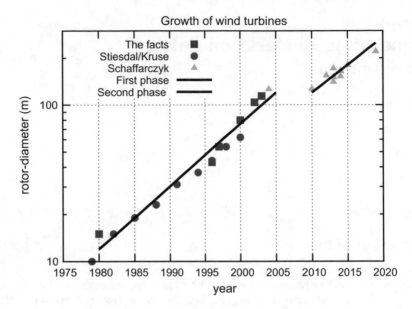

Fig. 10.1 Increase of sizes of wind turbine rotor diameters, exponential versus non exponential size growth

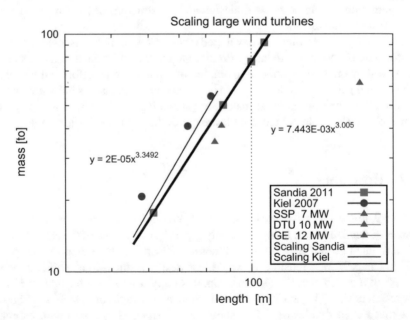

Fig. 10.2 Scaling of wind turbine blade masses (in metric tons)—Large Turbines only. Mass of GE's Haliade XL is estimated and shows a big technological leap to much smaller masses than expected

2011 ([27] in Chap. 9). A blade length of about 126 m was estimated accompanied with a blade mass of about 150 metric tons. One lesson learned is that standard aerodynamics seems to be insufficient for further upscaling. Therefore two further projects have been launched:

1. **INN**ovative **WIND** conversion systems (10–20 MW) for offshore applications (duration: November 2012–October 2017)
2. **Ad**Vanced **A**erodynamic **T**ools for l**A**rge **R**otors (duration: November 2013–October 2017). This project is administrated by the *European Energy Research Alliance*.

Somewhat different but nevertheless closely related to improve rotor performance, a German project *Smart Blades* (duration: December 2012 to February 2016) which concentrates on active flow control, is underway.

10.3 Outlook

Future developments of wind turbine aerodynamics are currently discussed more in terms of upscaling and less in terms of new aerodynamic concepts. Increasing the size immediately gives increased masses and one begins to wonder if (normalized) prices will change. We have emphasized this in the last sections. As energy eco-

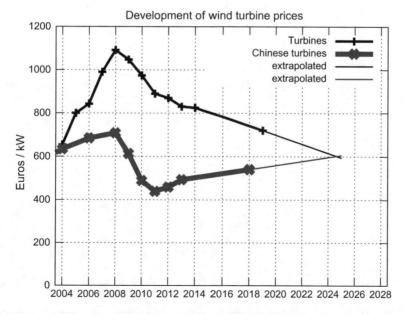

Fig. 10.3 Development of wind turbine prices, *Sources* Wind power monthly, IEA technology road map wind energy, [6] CWEA, Wang ZhongXia (prov.com) and others

nomics is closely connected to energy policy [5], we may see the driving force for further aerodynamic research from this point of view: the price per kWh of produced electricity [4][1] must be decreased further.

One important part of overall cost comes from investment in the turbine. Figure 10.3 shows the development for turbines in general and Chinese [6] turbines in particular. It is interesting to see that in 2004 both prices were equal and reached a maximum (at different levels) in 2008. Since 2012 a trend toward convergence (in 2025) may be seen.

References

1. Bak C et al (2013) The DTU 10-MW reference wind turbine, Danish Wind Power Research 2013
2. Siros G et al (2012) UPscaling wind turbines: theoretical and practical aspects and their impact on the cost of energy. Wind Energy 15:3–17
3. Chaviaropoulos P, Siros G (2014) Design of low induction rotors for use in large offshore wind farms. In: Proceedings of the EWEA 2014 annual event, Barcelona, Spain
4. Lantz E et al (2012) IEA wind task 26 the past and future cost of wind energy, NREL/TP-6A20-53510. Golden, CO, USA
5. NN, World Energy Outlook (2013) IEA/OECD. France, Paris, p 2013
6. NN, IEA WIND 2012 annual report, IEA WIND, ISBN 0-9786383-7-9, July 2013
7. Peeringa J, Brood R, Ceyhan O, Engles W, de Winkel G (2011) Upwind 20MW wind turbine pre-design, ECN-E-11-017. Petten, The Netherlands

---

[1] It is interesting to observe [5] the difference for residential electricity prices around the world from 2012 to 2035 (in US$ per MWh): Japan: 240, European Union 200, USA 120–130 and 70–90 for China. Only in China, prices for industrial consumers are higher than for residential consumers.

Appendix
Solutions

A.1 Solutions for Problems of Sect. 1.2

1.1 Power density in the wind:

   6 m/s: 136 W/m^2
   8 m/s: 323 W/m^2
   10 m/s: 630 W/m^2
   12 m/s: 1089 W/m^2
   Annual averaged values:
   4 m/s: 46 W/m^2
   5 m/s: 89 W/m^2
   6 m/s: 154 W/m^2
   7 m/s: 245 W/m^2

1.2

   According to [4] in Chap. 1 Eq. (7) we have (*1-2-3-formula*)

$$\bar{P} = \rho^1 \left(\frac{2}{3}D\right)^2 \cdot \bar{v}^3 = 1.13 \cdot \frac{\rho}{2}\frac{\pi}{4}\bar{v} \tag{A.1}$$

for a Betz turbine ($c_P = 16/27 = 0.596$) (Table A.1).

1.3 Let m = 0. Then

$$p(y,\sigma) = \frac{1}{2\pi\sigma^2}\int du \int dv \cdot exp(-u^2/2\sigma^2) \cdot exp(-v^2/2\sigma^2) \cdot \delta(y - \sqrt{u^2 + v^2}). \tag{A.2}$$

Introducing 2D polar coordinates $u = r \cdot cos(\varphi)$ and $v = r \cdot sin(\varphi)$, it follows

$$p(y,\sigma) = \tag{A.3}$$

$$\frac{1}{2\pi\sigma^2}\int_0^{2\pi} d\phi \int_0^\infty dr \cdot r \cdot exp(-r^2/2\sigma^2)\delta(r - y) == \tag{A.4}$$

© Springer Nature Switzerland AG 2020
A. P. Schaffarczyk, *Introduction to Wind Turbine Aerodynamics*,
Green Energy and Technology, https://doi.org/10.1007/978-3-030-41028-5

Table A.1 Numerical values

c_P	$\bar{c}_P$
16/27	1.13
0.58	1.10
0.5	0.948
0.4	0.758

$$\frac{y}{\sigma^2} \cdot exp(-y^2/2\sigma^2). \qquad (A.5)$$

1.4 $\eta = 6/30 = 0.186$.

A.2 Solutions for Problems of Sect. 2.10

2.1 Calculating $\frac{\partial c_P}{\partial \lambda} = 0$ we arrive at

$$\lambda_m^2 - \frac{2}{3}\frac{c_2}{c_1} \cdot \lambda - \frac{1}{3} = 0 \qquad (A.6)$$

with the solution

$$\lambda_m = \frac{1}{3}\left(\frac{c_2}{c1} \pm \sqrt{\left(\frac{c_2}{c_1}\right)^2 - 3} \right). \qquad (A.7)$$

2.2 Power increases with swept area which is $\sim D^2$. Therefore

$$P = P_= \cdot \left(\frac{A_D}{A_R}\right) \cdot \eta = 1.16 \qquad (A.8)$$

In practice and especially for a small wind turbine, losses have to be included. They may vary between 20 and 50%.

2.3 Clearly two rotors have to be present but they are extracting energy from the same stream tube. In the best case then these two rotors feed in torque on one shaft. See Fig. A.1 as an example.

A.3 Solutions for Problems of Sect. 3.11

3.1 The Euler equation for steady flow reads as

$$\rho(\mathbf{u} \cdot \nabla)\mathbf{u} = -\nabla p\mathbf{f} . \qquad (A.9)$$

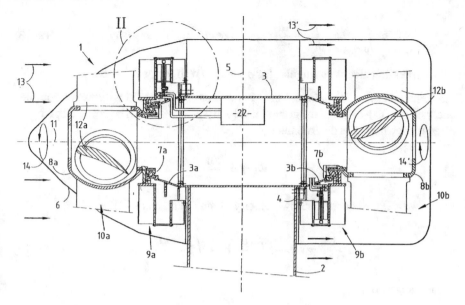

Fig. A.1 Example of a drive train for a counter-rotating wind turbine

Assuming that $\mathbf{f} = \nabla\Phi$[1] and using $\mathbf{u} \cdot \nabla\mathbf{u} = \nabla(\frac{1}{2}\mathbf{u} \cdot \mathbf{u}) - \mathbf{u} \times \omega$, we obtain

$$-\mathbf{u} \times \omega = -\nabla\left(\frac{p}{\rho} + \Phi + \frac{1}{2}\mathbf{u}^2\right) \tag{A.10}$$

If there is no vorticity $\omega = 0$ then *along* a streamline $\mathbf{u} = \frac{d\mathbf{r}}{dt}$, the quantity H

$$H := \frac{p}{\rho} + \Phi + \frac{1}{2}\mathbf{u}^2 \tag{A.11}$$

remains constant.

Remark Read [4] in Chap. 3 for a careful discussion on the abuse of *Bernoulli's* equation for explaining lift.

3.2 A line vortex with concentrated constant circulation along the z-axis (x = y = 0) induces a purely tangential velocity field $\mathbf{u}_\phi$,

$$\mathbf{u}_\phi = \frac{\Gamma}{2\pi r}. \tag{A.12}$$

As kinetic energy in a given volume V is $E = \rho \int_V v^2 \cdot dV$, we have per unit length of vortex line:

---

[1] van Kuik [9] argues that for an actuator disk this may not be generally true.

$$E = \int_0^{2\pi} d\phi \cdot \int_\varepsilon^R \frac{r \cdot dr}{r^2} = const \cdot (log(R) - log(\varepsilon)) \ . \tag{A.13}$$

This expression clearly is singular for $\varepsilon \to 0$ as well as for $R \to \infty$.

3.3 Remember that 2D potential flow is most easily discussed with use of complex analysis $z = x + i \cdot y$ and the complex velocity $w = u - i \cdot v$. The complex potential $F(z)$ form where w is determined by

$$w(z) = \frac{dw}{dz} \tag{A.14}$$

reads (with U the constant inflow velocity along the real axis [2]) as

$$\frac{1}{U} F(z) = -i \left(\frac{kU}{w}\right)^{\frac{1}{2}} + \left(1 - \frac{kU}{w}\right)^{\frac{1}{2}} . \tag{A.15}$$

Upon integration:

$$2i \left(\frac{w}{kU}\right)^{\frac{1}{2}} + \left(\frac{w}{kU}\right)^{\frac{1}{2}} \left(\frac{w}{kU} - 1\right)^{\frac{1}{2}} + \frac{1}{2}\pi i - log \left\{\left(\frac{w}{kU}\right)^{\frac{1}{2}} + \left(\frac{w}{kU} - 1\right)^{\frac{1}{2}}\right\} = \frac{z}{k} , \tag{A.16}$$

with

$$k = \frac{2a}{\pi + 4} \text{ and } 2a = \text{ plate breadth.} \tag{A.17}$$

The flow is symmetrical along the real axis $y \to -y$. At point B $(z_B = a \cdot i) w = kU$. Let s be the distance from B on the boundary streamline, and [2] finds:

$$x(s) = (s^2 + sk)^{\frac{1}{2}} - k \cdot log \left\{\left(\frac{s}{k} + 1\right)^{\frac{1}{2}} + \left(\frac{s}{k}\right)^{\frac{1}{2}}\right\} , \tag{A.18}$$

$$y(s) = 2(sk + k^2)^{\frac{1}{2}} + \frac{1}{2}\pi k. \tag{A.19}$$

For $s \to \infty$ the last terms in Eqs. (A.18) and (A.19) may be dropped and s can be eliminated so that the asymptotic shape

$$y^2 = 4kx = \frac{8a}{4 + \pi} x \tag{A.20}$$

is an quadratic parabola.

3.4 The (conformal) mapping between the $\zeta = \xi + i \cdot \eta$ and $z = x + i \cdot y$ planes is given by

$$z = \zeta + \frac{\lambda^2}{\zeta} \tag{A.21}$$

and the subset (a circle) is due to the equation

Table A.2 Results

ε	$u'^2 (m^2/s^2)$	η (mm)	Taylor microscale (mm)
$5.0 \cdot 10^{-3}$	1.00	0.743	34.5
$4.5 \cdot 10^{-2}$	2.25	0.429	8.6
$3.0 \cdot 10^{-2}$	4.00	0.475	23.0

$$\zeta = \zeta_m + R \cdot e^{i \cdot \varphi} \quad 0 \leq \varphi < 2\pi \tag{A.22}$$

$$z_m = x_m + y_m. \tag{A.23}$$

R is determined by the constraint that the circle must pass through $z = 1$. If $z_M = 0$ and $\lambda = 1$, Eq. (A.21) maps to a strip on the real axis with $-2 < x < 2$ so that the chord is $c = 4$. For other values for λ and z_M, we have [2]

$$c \approx 4\lambda \left\{ 1 + \left(\frac{x_M - \lambda}{\lambda} \right) \right\}. \tag{A.24}$$

The (relative) thickness is approximately given by

$$t \approx \frac{3}{4} \sqrt{3} \cdot \left(\frac{x_M}{\lambda} - 1 \right). \tag{A.25}$$

Shifting z_M off the real axis introduces *camber* expressed through an angle β

$$y_M = \lambda \cdot \tan(\beta) \tag{A.26}$$

so that for zero angle of attack, a finite lift is generated:

$$c_{L,0} = 2\pi \cdot \tan(\beta). \tag{A.27}$$

3.5 (a) Kolmogorov and Taylor scales (Table A.2).

Note that all values have an estimated uncertainty of 10% at least.

(b) Spatial average and resolution of sensors.

It is clear from its definition Eq. (3.109) that the spatial correlation (or if we use Taylor's *frozen turbulence* hypothesis temporal as well) is estimated by λ_T. If we use of an unavoidable finite sensor (of characteristic length scale ℓ) typical correction-factors must be of order

$$constant \cdot \left(\frac{\ell}{\lambda_T} \right). \tag{A.28}$$

It can be estimated that if the sensor length is of the same size as Taylor's length, then an error of about 10% may be introduced. See more, e.g., in [14] for the case of hot-wire sensor in boundary layer flow.

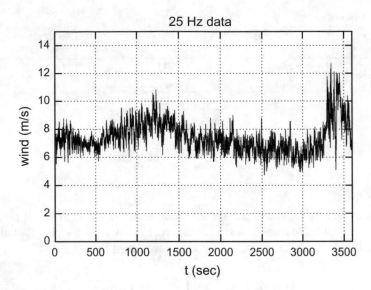

Fig. A.2 Time series of 1 h sample wind data

A.4   Solutions for Problems of Sect. 4.7

4.1 Influence of wind shear:
  (a) $z_0 = 1$ mm: $v_{avr} = 11.87$ m/s, $z_0 = 10$ mm: $v_{avr} = 11.83$ m/s.
  (b) $z_0 = 1$ mm: $P_{avr}/P = 0.974$, $z_0 = 10$ mm: $P_{avr}/P = 0.97$.

4.2 TI correction for power:
  To start we present a solution for cubic dependence $P(v) \sim v^3$:

$$P(v) = \frac{\rho}{2} c_P \frac{\pi}{4} v^3, \tag{A.29}$$

$$v(t) = \bar{v} + v', \quad \bar{v} = <v>, \quad <v'> = 0, \tag{A.30}$$

$$< (\bar{v} + v')^3 > = \bar{v}^3 + 3\bar{v}^2 \cdot <v'> + 3\bar{v} \cdot <v'^2> + <v'^3> = \tag{A.31}$$

$$= \bar{v}^3 + 3\bar{v}\sigma. \tag{A.32}$$

For more general dependencies expand P(v) into a power series:

$$f(x - x_0) = \sum_{n=1}^{\infty} \frac{f^{(n)}(x_0)}{n!} (x - x_0)^n. \tag{A.33}$$

4.3 Analysis of wind data (Figs. A.2 and A.3):
  (a) Gaussian distribution
  (b) Energy spectra (Fig. A.4)
  (c) Correlation time

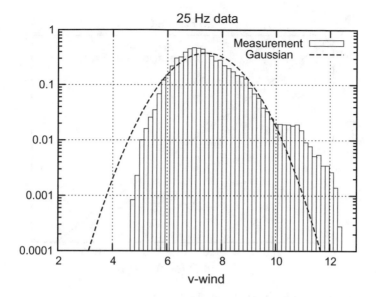

Fig. A.3 Histogram of 1 h wind data equivalent to 90 k data

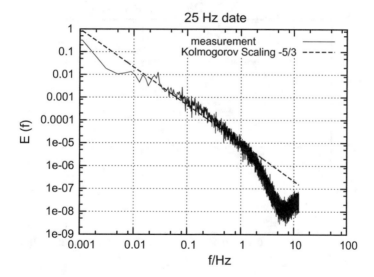

Fig. A.4 Energy content

Correlations often decrease exponentially

$$c(t) = exp-(t/T). \tag{A.34}$$

This cannot be seen in Fig. A.5, even not the decrease to $1/e = 0.368$ (Fig. A.6).

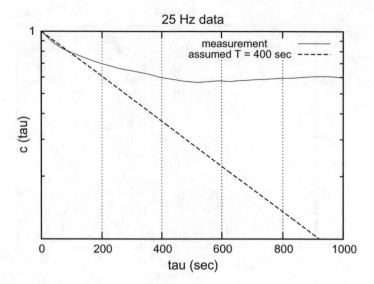

Fig. A.5 Correlation function and exponential with T = 400 s

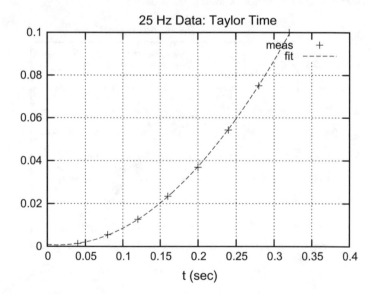

Fig. A.6 Quadratic fit according to [15] gives a Taylor time of about 30 ms

According to Eq. 3.109 and [15], the part of c(t) close to $t \approx 0$ can be deduced from a quadratic fit: $c(t) = a \cdot t^2 + b \cdot t + c$. From this $\lambda_T \approx \sqrt{c}$ a value may be deduced.

Table A.3 Results for Gaussian distribution

T in years	U_{max}
1	12.3
10	12.6
50	12.9
100	13.0

Table A.4 Results for a Rayleigh distribution

v (m/s)	$10^6 \cdot r(v)$	Return time (years)
22	24.9	0.76
23	9.76	1.95
24	3.65	5.21
25	1.31	14.6
26	0.45	42.6
27	0.15	130

4.4 Model of gust return time:

(a) Because [2] assumes Gaussian distributions for U and its derivatives, the values are comparably small (Table A.3).

(b) A Rayleigh distribution which is often used for a first guess gives (Table A.4).

4.5 For simplicity we assume that a gust may be described by constant wind speed of value u_g during a period T_g. If there is only one such gust during the measurement (and averaging) time T_m, the averaged wind speed should vary only linearity with T. This then gives $v_{3s} \approx 90$ m/s. An extrapolation along the lines of [2] gives a much higher values of 220 m/s, see Fig. A.7.

4.6 (a)TurbSim

The results were generated with the standard Kaimal.inp file but with a somewhat higher resolution: $\Delta t = 0.01$ s and $T = 600$ s (Figs. A.9 and A.10).

(b) Mann Model

These result come from a time series of the main flow component (Figs. A.12 and A.13).

*(c) CTRW

Here we present a sample (Fig. A.15).

A.5 Solutions for Problems of Sect. 5.7

5.1 Wind turbine used as a fan:

(a) Simple momentum theory applied to a fan [5] in Chap. 5

$$T_{ideal} = \sqrt[3]{2\rho \cdot A_r P^2}. \tag{A.35}$$

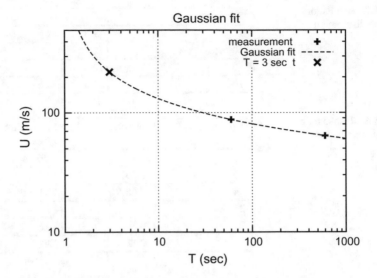

Fig. A.7 Extrapolation of 3 s gust wind speed assuming Gaussian distribution

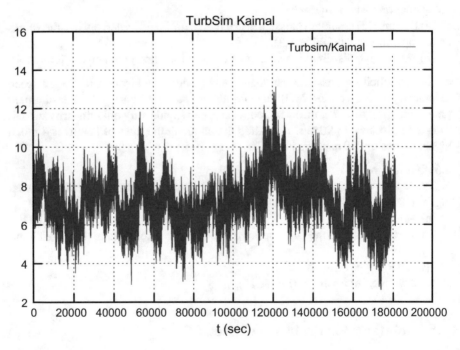

Fig. A.8 Sample time series from TurbSim

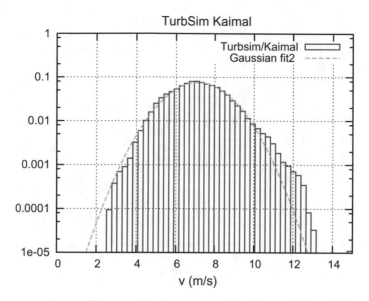

Fig. A.9 Sample histogram for time series from A.8

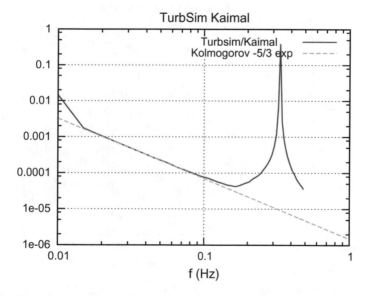

Fig. A.10 Sample energy spectrum for time series from A.8

As first inputs, we use our rated values $P = 330$ kW and 45 RPM.
Then $T_{fan} = 61$ kN, $v_2 = 5.4$ m/s, $v_3 = 2 \cdot v_2 = 11$ m/s.
Now lowering P down to 200 kW, the corresponding numbers become
$T_{fan} = 40{,}7$ kN, $v_2 = 4.9$ m/s, $v_3 = 2 \cdot v_2 = 9.8$ m/s.

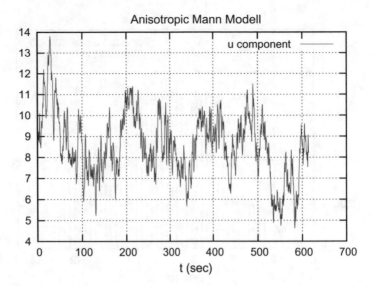

Fig. A.11 Sample time series from anisotropic Mann Model [7]

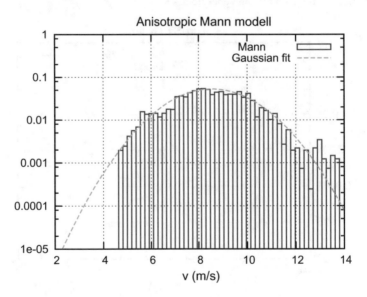

Fig. A.12 Sample histogram for time series from A.11

(b) BEM model

If we assume that NREL's wt-perf is reliable for the fan mode as well we get for the same RPM and a negligible inflow velocity of $v_1 = 0.01$ m/s the following values: $T_{fan} = 22$ kN, $P = 199$ kW, and $v_3 \approx 7$ m/s.

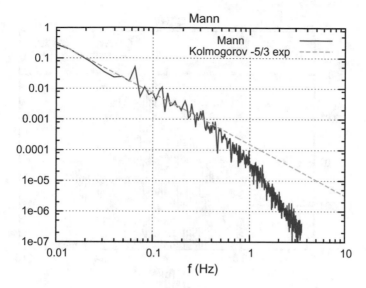

Fig. A.13 Sample energy spectrum for time series from A.11

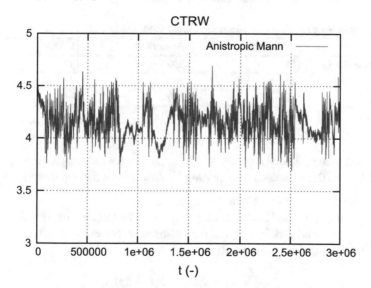

Fig. A.14 Sample time series from Kleinhans continuous time random walk model [14] in Chap. 4

Pitch was varied until (negative!) thrust was minimal.

Compared to pure momentum theory, this thrust is smaller by a factor of two. Therefore one may write his/her own code to have a third approach:

If we use the same RPM of 45 and the same pitch we now have

$T_{fan} = 38\,\text{kN}$, $P = 175\,\text{kW}$, and $v_3 = 8.5\,\text{m/s}$ (Table A.5).

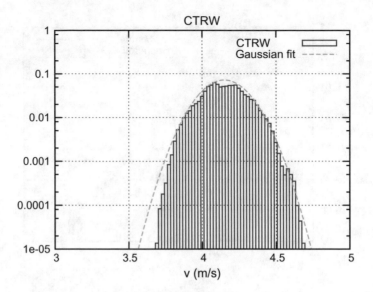

Fig. A.15 Sample histogram for time series from Fig. A.14

Table A.5 Summary of theoretical results for a wind turbine to be used a fan

Model	Thrust/kN	Power/KW	v_3/m/s
Momentum theory	61	330	11
Momentum theory	41	200	10
Wt-perf	22	199	~7
Own code	40	185	~9

Pure momentum theory neglects all losses so that we may expect less thrust for the BEM case. The final decision can only be drawn from the experiment.

5.2 Multiple actuator disks [20] in Chap. 5:

(a) Equations 5.65 and 5.66 are simply the BEM equations for two disks if the inflow for the second disk is the wind far downstream for the first disk.

(b) Using Eq. (5.64) we arrive at

$$c_{P2} = (1 - 2a)^3 \cdot 4b(1 - 2b)^3. \tag{A.36}$$

As there is no back reaction from disk 2 to disk 1, one finds for a maximum of disk 2:

$$b = 1/3, \tag{A.37}$$

$$c_{P2} = (1 - 2a)^3(16/27). \tag{A.38}$$

The total power of both disks then is a function of only

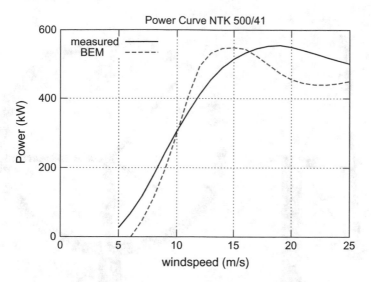

Fig. A.16 Comparison of measured data and a simple BEM model with global pitch of $-6°$

$$c_{P,totaL} = 4a(1-a)^2 + (1-2a)^3(16/27) . \tag{A.39}$$

By simple algebra one finds the results from Eqs. (5.67) and (5.70).

(c) A formal proof that Eq. (5.71) is correct is possible by induction.

The article further discusses different shapes for an array of two disks.

5.3 This turbine is discussed in more detail in [26] in Chap. 3 (pp 62 ff) and [38] in Chap. 7. Profile data has to added. If we choose from Abbot–Doenhof data from NACA 63 series, we get a picture as in Fig. A.16. Obviously this does not fit.

5.4 Wilson (Chap. 5 from [42]) uses *along* streamlines conservations laws of

$$\text{mass} : u_1 r_2 dr = u_3 r_3 dr, \tag{A.40}$$

$$\text{angular momentum} : r_2^2 \omega = r_3^2 \omega_1, \tag{A.41}$$

$$\text{and, energy} \frac{1}{2}(u_1 - u_2)^2 = \left(\frac{\Omega + \omega_1/2}{u_3} - \frac{\Omega + \omega/2}{u_2} \right) u_2 \omega_1 r_3^2. \tag{A.42}$$

In addition an equation *across* the streamlines is necessary which may be obtained from Euler's equation (3.31) in Chap. 3:

$$\frac{d}{dr_3} \left(\frac{1}{2}(u_1^2 - u_3^2) \right) = (\Omega + \omega_1)\frac{d}{dr_3}(\omega_1 r_3^2). \tag{A.43}$$

Axial inductions at the disk and far downstream are different:

$$u_2 = u_1 \cdot (1 - a) , \tag{A.44}$$

$$u_3 = u_1 \cdot (1 - b) . \tag{A.45}$$

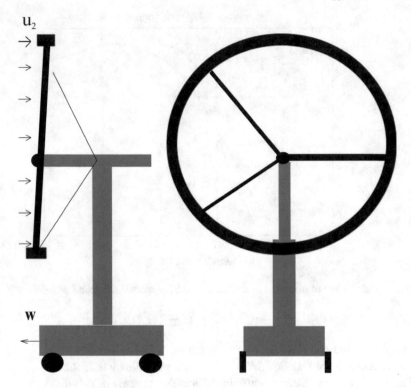

Fig. A.17 Principles of wind car

Using this and the energy equation Eq. (A.42) one gets

$$a = \frac{b}{2}\left[1 - \frac{(1-1)b^2}{4\lambda^2(b-a)}\right].$$ (A.46)

The interested reader should also compare [32, 35] in Chap. 5.

5.5 Wind Vehicle Part 1: Momentum theory:

Estimation of Gross Parameters

Figure A.17 shows the principle of a wind vehicle. Before we proceed to the formal analysis, it may be useful to present some typical numbers: at the races between university wind car teams the rotor area is limited to $2 \times 2 = 4\,\mathrm{m}^2$.

Using a HAWT the reduction goes further to $\pi = 3.14\,\mathrm{m}^2$. With a wind speed of $7\,\mathrm{m/s}$ and an assumed car speed of $3.5\,\mathrm{m/s}$, this gives—using an ordinary wind turbine—$P = 1320\,\mathrm{W}$ and $T = 188\,\mathrm{N}$. If no other forces have to be overcome, we need for translating the thrust alone $P_{thrust} = 660\,\mathrm{W}$. This is a highly ideal case. In reality a well-designed (SWT)rotor has only $c_P = 0.35$ that is only 60% of the power from above $= 780\,\mathrm{W}$. If the car has a roller friction of (only) 50 N, we have to have a net power of $50 \times 3.5 = 175\,\mathrm{W}$. Therefore this car does **not** drive against the wind.

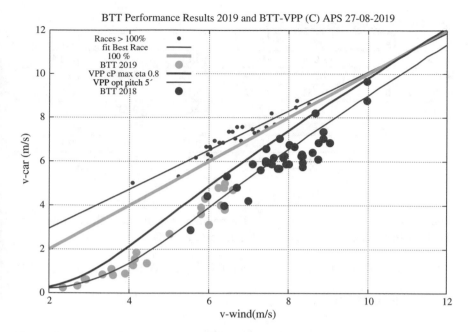

Fig. A.18 Results for the wind vehicle Baltic Thunder from 2010 to 2019. Compared to the first edition an almost doubling of performance (50% → 100% has been achieved)

By analyzing many races (see Fig. A.18) it is clear that $\tilde{w} = v_{car}/v_{wind} > 0.5$ was frequently reached (the actual *official* world record being 72%), so that a design goal may be $\tilde{w} \geq 0.5$.

Momentum Theory

Now, to derive a momentum formulation we follow again [40], in Chap. 5. To drive the car we need power of

$$P = (D + T) \cdot w \, . \tag{A.47}$$

On the other side, wind contains a power of

$$P = T \cdot u_1 \cdot (1 - a) \tag{A.48}$$

as usual. All other drag forces on the car are assumed to be summarized in

$$D = \frac{1}{2}\rho A_V (u_1 + w) \cdot c_D \, . \tag{A.49}$$

Equalizing Eqs. (A.47) and (A.48) we have

$$2\rho A_R(u_1 + w - u_2)u_2^2 = 2\rho A_R(u_1 + w - u_2)u_2 w + \frac{1}{2}\rho A_V c_D (u_1 + w)^2 w \, . \tag{A.50}$$

Table A.6 Results from solving Eq. (A.54)

K	a	v_c/v_{wind}	c_P	c_T	c_P/c_T	$\eta_{Drivetrain}$
0.013	0.05	2.0	0.16	0.18	0.89	>0.8
0.06	0.1	1.0	0.29	0.35	0.83	0.7
0.14	0.13	0.75	0.37	0.47	0.79	0.7
0.33	0.18	0.5	0.44	0.60	0.73	0.7
1.0	0.24	0.25	0.50	0.74	0.68	0.7
3.1	0.29	0.1	0.51	0.82	0.62	0.7
36	0.31	0.01	0.52	0.86	0.60	0.7

Note that now $u_2 = (u_1 + w) \cdot (1 - a)$. The non-dimensional car–velocity (the *ratio*) may be defined as $\tilde{w} := w/u_1$, so that Eq. (A.50) reads as

$$a(1 - a)^2 \cdot (1 + \tilde{w})^3 - a(1 - a) \cdot \tilde{w}(1 + \tilde{w})^2 - K \cdot \tilde{w}(1 + \tilde{w})^2 = 0, \quad \text{(A.51)}$$

with K from Eq. (5.78) and

$$\tilde{w} = \frac{a(1 - a)^2}{K + a^2(1 - a)} . \quad \text{(A.52)}$$

The car's speed is at a maximum if

$$\frac{d\tilde{w}}{da} = 0 , \quad \text{(A.53)}$$

which gives after some simple algebra the cubic equation:

$$a^3 - a^2 - 3ka + K = 0 . \quad \text{(A.54)}$$

Results

Equation A.54 may be solved by either using a numerical (Newton–Raphson) algorithm or directly by using Cardano's method. Results are show in Table A.6.

5.6 Derivation of Prandtl's tip correction.

2D Fluid Mechanics and Complex Analysis

2D potential flow is most elegantly discussed with use of complex analysis. Let $\mathbb{C} \ni z := x + i \cdot y, i^2 = -1$ and $(x, y) \in \mathbb{R}$. Any inviscid irrotational flow field then may be described by a stream function Φ (to obey the continuity equation) and a velocity potential Ψ $((u, v) = \nabla \cdot \Psi)$ which are combined into one complex quantity $w = u - i \cdot v$ and $F(z) = \Phi + i \cdot \Psi.^2$ F is called the *complex (velocity) potential*.

---

[2]Note that the complex potential and the tip correction factor have the same symbol F.

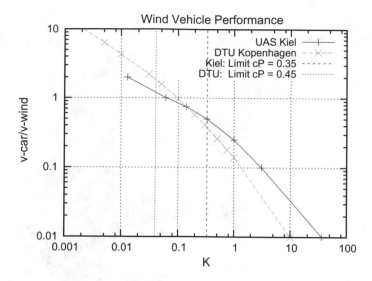

Fig. A.19 Plot of results for different rotor optimizations

Velocity is related to F via complex derivation like

$$w = \frac{dF}{dz} .$$

(A.55)

One special technique possible only in complex analysis is the use of *conformal mapping* [3] $f : \mathbb{C} \rightarrow \mathbb{C}$ of which the Joukovski map

$$f(z) = z + \frac{c^2}{z}$$

(A.56)

is the most well known. It maps a circle into a shape which has become famous as the *Joukovski profile* [1]. As a result of conformal mapping the simpler flow field around a cylinder is mapped by the same function to the more complicated flow around this profile.

Further, complex analysis shows that **real** multivalued functions like $\sqrt{(x)}$ or $\log(x)$ may be transformed to a one-to-one complex-valued function by introducing the concepts of *analytical continuation* and *Riemannian surfaces*. As an example, Fig. A.20 shows the complex logarithm but a proper picture in 3D is not possible because of Nash's embedding theorem. In general all manifolds are smooth and there is no intersection between the different *branches*. For a more mathematical introduction to this **extremely beautiful and easy** subject see [10].

Fig. A.20 Example of a
Riemannian surface: the
complex logarithm

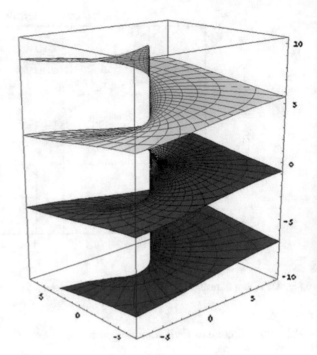

Model Description and Derivation

Prandtl starts with a simple circulatory flow on $\zeta \ni G \subset \mathbb{C}$; $|\zeta| > 1$ whose potential
is **assumed** to be

$$F(\zeta) = \Phi + i \cdot \Psi := ic \cdot log(\zeta) \tag{A.57}$$

with $\zeta = \xi + i \cdot \eta$ and $\xi, \eta \in \mathbb{R}$.

By using the map $z : \zeta \mapsto z$

$$z(\zeta) := \frac{d}{\pi} \cdot log\left(\frac{1}{2}\left(\zeta + \frac{1}{\zeta}\right)\right), \tag{A.58}$$

G is mapped into a sliced complex plane, see Fig. A.21. Each circle in the ζ-plane $\zeta = R \cdot e^{i \cdot \phi}$ with $R = R_0 + dr$ with $R_0 = 0.01$; $dr = 1.0$; $0 \le \phi < 2 \cdot \pi$ corresponds
to a wavy line and seems to be sliced by strips of distance $2 \cdot d$ in the imaginary
direction in the z-plane.

Using the definition of the complex cosine:

$$cos(z) := \frac{e^{iz} + e^{-iz}}{2}, \tag{A.59}$$

we have

$$z(F) = x + i \cdot y = \frac{d}{\pi} \cdot log\ cos\left(\frac{F}{c}\right). \tag{A.60}$$

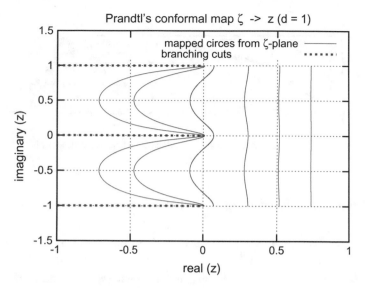

Fig. A.21 Potential flow around a stack of plates induced by a circulatory flow and a conformal mapping

Discussing the streamline $\Psi = 0$, F becomes real and y has to vanish, so

$$x = \frac{d}{\pi} \cdot log \, cos \left(\frac{\Phi}{c} \right) , \tag{A.61}$$

or

$$\pm \Phi = c \cdot \arccos(\exp(\pi x/d)) , \tag{A.62}$$

because $cos(\pm z) = cos(z)$, $\forall z \in \mathbb{C}$. From that Prandtl concluded that there is a jump in velocity potential from $x \to 0_-$ to $x \to 0_+$ by

$$\Delta\Phi = 2c \cdot arc \, cos(exp(\pi x/d)) . \tag{A.63}$$

Prandtl further equates the jump in potential to the change in circulation which may be justified because (here Δ means a small finite difference, not the Laplacian)

$$\Delta v = \nabla\Phi \cdot dr = (\Phi_+ - \Phi_-) \cdot dr \tag{A.64}$$

and

$$\Delta\Gamma = \Delta \int_C v \cdot dr = \Delta v \cdot dr . \tag{A.65}$$

A slightly different derivation is presented in [5]. Now identifying $-x := R_{tip} - r$, r being the distance from the hub $r \approx R_{tip}$, and demanding $\lim_{d \to 0} \Gamma \to 1$ we have

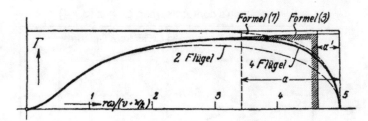

Fig. A.22 Modified circulation distribution at the edge of the actuator disk by Prandtl's correction from [2]

to normalize Φ (and F) according to $c = 1/\pi$ to arrive at

$$\Delta\Phi := F = \frac{2}{\pi} \cdot arc\ cos(exp(-\pi(R_{tip} - r)/d)) \ . \tag{A.66}$$

What remains is to determine d, the spacing of the stacks. As the derivation was an appendix of Betz' optimization problem for a ship propeller giving highest thrust for a prescribed power ($a < 0$ in wind turbine terms), Prandtl likens d to the distance of the helical vortex sheets which emerge from the actuator disk and which have to be rigid as a result of Betz' variational solution to his optimization problem. In its original form Prandtl gives for d to zeroth ($a \approx 0$) order $\lambda = \omega \cdot R/u_\infty$ the tip–speed ratio:

$$d_0/R_{tip} = \frac{2\pi}{B\lambda} \cdot \frac{\lambda}{\sqrt{1 + \lambda^2}} \ , \tag{A.67}$$

which may be improved [5, 17] by using the flow angle at the tip

$$tan(\varphi) = \frac{1 - a}{1 + a'} \cdot \frac{1}{\lambda} \tag{A.68}$$

(a and a' are the usual induction factors from wind turbine momentum theory in axial and tangential direction) to

$$d/R_{tip} = \frac{2\pi}{B} \cdot sin(\varphi) \ . \tag{A.69}$$

Putting everything together we finally have

$$F = \frac{2}{\pi} \cdot \arccos \exp\left(-\frac{B(R_{tip} - r)}{2R_{Tip} \cdot sin(\varphi)}\right) \ . \tag{A.70}$$

In Fig. A.22 we show the original picture from [2] (Flügel = blade, Formel = equation).

Discussion and Extension to BEM

Obviously there is a lot of arbitrariness in the derivation:

- The conformal map Eq. (A.58) is by no means justified by fluid mechanics of propellers.
- The dependency on the number of blades is an assumption as well.[3]

Glauert [5] gives a summary of Prandtl's and Goldstein's derivations suggesting the improvement of Eq. (A.68) for use of the tip-correction (better: finite number of blades)in the Blade-Element-Momentum (BEM) method.

He introduced (without any detailed reasoning) a *first-order correction*

$$\frac{a}{1+a} = \frac{\sigma c_{tan} \cdot F}{4 \cdot sin^2\varphi}, \tag{A.71}$$

$$\frac{a'}{1-a'} = \frac{\sigma c_{nor} \cdot F}{4 \cdot sin\varphi \cdot cos\varphi}. \tag{A.72}$$

In the later classical reviews of de Vries [44] in Chap. 5 and Wilson, Lissaman and Walker [45] in Chap. 5 ad-hoc *higher corrections* are introduced, meaning that some a, a' are replaced by $a \cdot F, a \cdot F'$.

A thorough discussion of the physical implication (and inconsistencies) was given in [17] with the result that a new model is proposed:

$$F_{new} = \frac{2}{\pi} \cdot arccos\left[exp\left(-g\frac{B(R_{tip} - r)}{2R_{tip}sin(\varphi_{r_{tip}})}\right)\right], \tag{A.73}$$

$$g = exp(-c_1(B\lambda - c_2)). \tag{A.74}$$

Application to the US NREL Phase VI (around 2000) and Swedish–Chinese FFA-CARDC (around 1990) measurements showed much better predictions of the loading in the tip region.

Comparison to Other Approaches

Sidney Goldstein, at that time a Ph.D. student of Prandtl, solved the problem of finding a (potential-theoretic) flow field for a finite number of blades but for lightly loaded propellers only. The accurate numerical evaluation is regarded to be complicated and was under discussion until recently, see [13]. Figure A.23 gives a comparison for $B = 3$ and λ (TSR) $= 6$.

With the emergence of CFD, Hansen and Johansen [8] investigated the flow close to wind turbine tips using this method. They found strong dependencies on the specific tip shape and confirmed some of the results from [17].

---

[3]In particular it is hard to imagine how a one-bladed rotor is pancaked to a whole disk homogeneously.

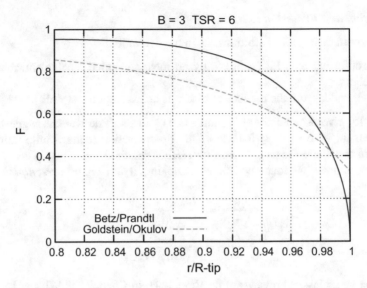

Fig. A.23 Comparison of Prandtl's and Goldstein's optimal circulation distribution

Summary

Prandtl's tip-correction model applies to the drop in circulation at the tip of a rotor and is now almost 100 years in use. Its derivation is remarkably simple and relies only on simple but ad-hoc conformal mapping. Its use in BEM codes is essential.

It has to be noted, however that any generalized use of the original Eq. (A.70) in BEM is ad-hoc as well. Semi-analytical vortex methods like in [4] seem to guide the way for improved, well-founded and *simple* tip-correction models.
5.7 BEM model of ideal wind turbine (Figs. A.15 and A.24).

A.6 Solutions for Problems of Sect. 6.8

6.1 Wilson's [45] in Chap. 5 aerodynamic model for a Darrieus rotor:

Averaged Induction

Using Kutta–Joukovski we start with

$$\Gamma = \pi c W \cdot sin(\alpha). \tag{A.75}$$

Looking on the velocity triangle of Fig. 6.16 (right side) the projected forces (per unit height) from lift only are

$$\mathbf{F} = \rho \pi c \cdot \begin{pmatrix} -v_a v_t cos^2(\theta) \\ v_a cos(\theta) + v_a v_t cos(\theta) sin(\theta) \end{pmatrix}. \tag{A.76}$$

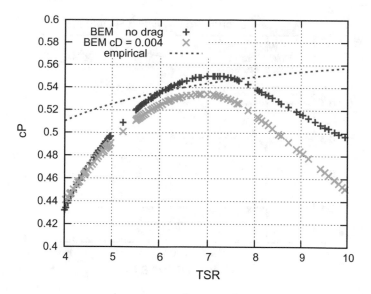

Fig. A.24 Estimation of performance of an ideal rotor ($B = 3$, $\lambda_{des} = 7$, $c_{L,des} = 1.0$)

The stream tube which is varying with rotation is

$$dy = R \cdot |sin(\theta)| \cdot d\theta . \tag{A.77}$$

One revolution with Ω will occupy the time $2\pi/\Omega$ in this stream tube. Force component from lift and momentum balance are

$$F_x = -2\rho\pi c v_a v_t cos^2(\theta)\frac{d\theta}{\Omega} = \tag{A.78}$$

$$= \rho\frac{2\pi}{\Omega}a(1-a)u_1 2u_1 R|cos(\theta)|d\theta. \tag{A.79}$$

From that we get Eq. 6.55.

Power Coefficient

For power we need torque and from that tangential forces. Equation (6.55) gives force in x(= wind = thrust) direction. Projecting into circumferential direction

$$Q = \rho\pi c R u_1^2(1-a)^2 \cdot sin^2(\theta) . \tag{A.80}$$

Inserting Eq. 6.55 and averaging over one revolution, one gets Eqs. 6.56–6.58.

Maximum Power

Finding the maximum is just simply to derivate Eq. 6.56 to x. the result should be $c_{P,max} = 0.554$ at $x_{max} = 0.802$.

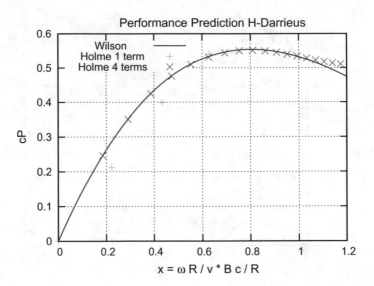

Fig. A.25 Graphical comparison for Power prediction

Thrust

We then insert Eq. 6.55 into Eq. A.79 to get

$$c_T(x) = \frac{x}{6} \cdot (3\pi - 4x) \ . \tag{A.81}$$

Discussion

$c_{P,max}$ here is smaller than Betz' limit and from double actuator disk, Problem 5.2. Nevertheless one important point is to have an indication of TSR to be chosen. For maximum power output:

$$\lambda_{opt} = 0.8/\sigma \ . \tag{A.82}$$

6.2 Solution for problem: Holme's [13] in Chap. 6 aerodynamical model for a Darrieus turbine. Iterative derivation of the series is not simple. Readers should use the attached FORTRAN code to run. Figures A.25 and A.26 show results for power and thrust. Notice how close the results are! Remember that both models break down for $x \lesssim 1$.

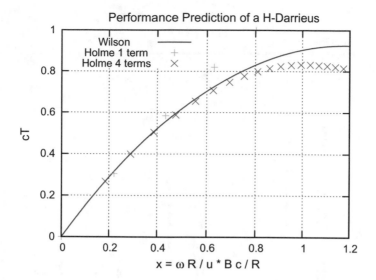

Fig. A.26 Graphical comparison for thrust prediction

FORTRAN Source Code of Holme's Linear Approach.

```
c     nach holme 1976
c
c     redux aps  2005 ... 2013
c
      program holme
c
      nmax = 8
c
      parameter(nmax=8)
      real mat(nmax,nmax),  b(nmax)
      real aa(0:nmax,nmax), bb(0:nmax,nmax)
      real la, mu, sig, ko, lamax, lamin
c
      pi = 4.*atan(1.)
      gra = 180./pi
      sig = 0.05
      write(*,101)'sig=',4.*sig
c
      open(unit=10,file='hol.dat',form='formatted',status='unknown')
c
      nl = 20
      lamin = 0.05
      lamax = 0.7
      dl = (lamax-lamin)/nl
c
      write(10,105)'lam','cp','cx','cy','cmdot','la*sig','sig*mu/0.42'
c
      do  ll=0,nl
      la = lamin+ll*dl
c
      write(*,*)'*************************************************'
      write(*,101)'la=',la
      ga = 1.+ la**2 + sig**2
c
      do i=1,nmax
        do j =1,nmax
        mat(i,j) = 0.0
        aa (i,j) = 0.0
        bb (i,j) = 0.0
        end do
        b(i)    = 0.0
      end do
c
c     hier die aa und bb besetzen
c
c     anzahl der gleichungen
c
      ib = nmax/2
      iend = ib/2
      write(*,*)'nmax ib iend',nmax, ib, iend
c
c     nur rechte seite der ersten gleichung = 1
c
      b(1) = 1.0
c
c     definiton of Holme's A
c
      do n=0,ib
      do m=1,iend
      ind = 2*m-1
      ko = float(ind*(-1)**m)/float((n**2-(ind**2)))
```

```
      aa(n,m) = (-1.)**(n/2)*4.*ko*la/pi
      end do
      end do
c
c     B
c
      do n=0,ib
      do m=1,iend
      ind         = 2*m
      ko   = float(ind*(-1)**m)/float((n**2-(ind**2)))
      bb(n,m) = (-1)**((n-1)/2)*ko*4.*la/pi
      end do
      end do
c
c     die ersten 3 gleichungen (7.6)
c-------------------------------------------
c     erste reihe
c
      do n=0,2,2
        do m=1,iend
        mat(1,m) =  mat(1,m) + aa(n,m)
        end do
      end do
      mat(1,1)    = ga + mat(1,1)
      mat(1,ib+2) = la*sig
c
c     zweite reihe
c
      mat(2,1)   = -la*sig
      mat(2,3)   = -la*sig
c
      do n=1,3,2
        do m=1,iend
        mat(2,ib+m) =  mat(2,ib+m) + bb(n,m)
        end do
      end do
c
      mat(2,ib+2) = ga + mat(2,ib+2)
c
c     dritte reihe
c
      mat(3,ib+2) = -la*sig
      mat(3,ib+4) = la*sig
c
      do n=2,4,2
        do m=1,iend
        mat(3,m) =  mat(3,m) + aa(n,m)
        end do
      end do
      mat(3,3)    = ga + mat(3,3)
c
c     vierte reihe
c
      mat(4,3)   = la*sig
      mat(4,5)   = -la*sig
c
      do n=3,5,2
        do m=1,iend
        mat(4,ib+m) =  mat(4,ib+m) + bb(n,m)
        end do
      end do
```

```fortran
      if(psi.lt.psimin)psimin=psi
      write(*,102)gra*tet,psi
      end do
      write(*,101)'psimin=',psimin
      write(*,101)'psimax=',psimax
c
      cmdot = 0.5*(psimax-psimin)
      write(*,101)'cmdot=',cmdot
      mu = cmdot*la/sig
      write(*,101)'(TSR)mu=',mu
      write(*,101)'sig*mu=',4.*sig*mu
      write(*,*)
c
      cp = 0.0
      do i=1,nmax
         cp = cp + b(i)**2
      end do
      cp = 2.*pi*mu*sig*cp
      write(*,101)'cp=',cp
c
      cx = 2.*pi*mu*sig*b(1)
      write(*,101)'cx=',cx
c
      cy = 0.0
      do n=1,ib
         cy = cy + b(n)**2
         j=ib+n
         cy = cy + b(j)**2
      end do
      cy = 2.*pi*sig*(mu*b(ib+1)-la*cy)
      write(*,101)'cy=',cy
c
      so =4.*sig*mu
      write(10,104)mu,cp,cx,cy,cmdot,so,so/0.42
c
      close(unit=10)
c
100   format(6f12.6)
101   format(a20,f10.4)
102   format(2f12.4)
103   format(2a12)
104   format(7f12.4)
105   format(7a12)
c
      end
c----------------------------------------
      subroutine gauss(a,b,n)
c
      integer n
      real a(n,n), b(n)
      integer icol,irow,j,k,l,ll
      integer ipiv(n),indc(n),indr(n)
c
      real big, dum, pivinv
      do j=1,n
         ipiv(j) = 0
      end do
c
```

```fortran
c
      mat(4,ib+4) =  ga + mat(4,ib+4)
c
c------------------------------------------------------
c     reihe > nmax/2 die b's nach holme  (7.7)
c
c row=ib+1
      mat(ib+1,1)    = sig
      mat(ib+1,ib+1) = -1.
c
c row=ib+2
      mat(ib+2,1)    = la
      mat(ib+2,2)    = 1.
      mat(ib+2,ib+2) = sig
c
c row=ib+3
      mat(ib+3,3)    = -sig
      mat(ib+3,ib+2) = la
      mat(ib+3,ib+3) = 1
c
c row=ib+4
      mat(ib+4,3)    = la
      mat(ib+4,4)    = 1.
      mat(ib+4,ib+4) = sig
c
      call gauss(mat,b,nmax)
c
c-------------- debug -------------
c
c     b(1) = 1./(1+1a*+2*sig*+2+16.*la/(3.*pi))
c     do j=2,ib
c        b(j) = -la*b(j-1)
c     end do
c     do j=ib+1,nmax
c        b(j) = 0.0
c     end do
c
c-------------- debug -------------
c
      write(*,100)((mat(i,j), j=1,nmax),i=1,nmax)
      write(*,*)'a:'
      write(*,100)( b(i), i=1,nmax/2)
      write(*,*)'b:'
      write(*,100)( b(i), i=1+nmax/2,nmax)
      write(*,*)
c
      teta = -0.75*pi
      tete = 0.75*pi
      nt = 100
      dt = (tete-teta)/nt
      write(*,103)'theta ','psi'
      psimax = -1.e3
      psimin = 1.e3
      do i=0,nt
         tet = teta+i*dt
         psi = 0.
      do j=1,ib
         psi = psi + (b(j)*sin(j*tet))/j
         js = ib + j
         psi = psi - (b(js)*cos(j*tet))/j
      end do
      if(psi.gt.psimax)psimax=psi
```

```
      do i=1,n
        big = 0.
        do j=1,n
          if(ipiv(j).ne.1)then
            do k = 1,n
              if(ipiv(k).eq.0)then
                if(abs(a(j,k)).ge.big)then
                  big=abs(a(j,k))
                  irow=j
                  icol=k
                endif
              else if(ipiv(k).gt.1)then
                write (*,*) 'singular matrix 1'
              end if
            end do
          end if
        end do
        ipiv(icol)=ipiv(icol)+1
        if(irow.ne.icol) then
          do l=1,n
            dum = a(irow,l)
            a(irow,l)=a(icol,l)
            a(icol,l)=dum
          end do
          dum      = b(irow)
          b(irow)= b(icol)
          b(icol)= dum
        end if
        indr(i)=irow
        indc(i)=icol
        if (a(icol,icol).eq.0.) write(*,*) 'singular matrix 2'
        pivinv=1./a(icol,icol)
        a(icol,icol)=1.
        do l=1,n
          a(icol,l)=a(icol,l)*pivinv
        end do
        b(icol)=b(icol)*pivinv

        do ll=1,n
          if(ll.ne.icol)then
            dum=a(ll,icol)
            a(ll,icol)= 0.
            do l=1,n
              a(ll,l)=a(ll,l)-a(icol,l)*dum
            end do
            b(ll)=b(ll)-b(icol)*dum
          end if
        end do
      end do
      do l=n,1,-1
        if(indr(l).ne.indc(l))then
          do k=1,n
            dum=a(k,indr(l))
            a(k,indr(l))=a(k,indc(l))
            a(k,indc(l))=dum
          end do
        end if
      end do
      return
      end
```

6.3 Scully's vortex:

$\boldsymbol{\omega} = \nabla \times \mathbf{v}$ in polar coordinates (r, ϕ, z) is

$$\begin{pmatrix} \dfrac{1}{r}\dfrac{\partial u_z}{\partial \phi} - \dfrac{\partial u_\phi}{\partial z} \\ \dfrac{\partial u_\phi}{\partial z} - \dfrac{\partial u_z}{\partial u_\phi} \\ \dfrac{1}{r}\left(\dfrac{\partial r u_\phi}{\partial r} - \dfrac{\partial u_\phi}{\partial \phi} \right) \end{pmatrix}. \tag{A.83}$$

The only component left is

$$\omega_z = 2\left(1 + r^{2n}\right)^{-\left(\frac{n+1}{n}\right)}. \tag{A.84}$$

As one may easily see, this is a strictly monotonic decreasing function for $r > 0$.

6.4 Hill's spherical vortex:
Readers are urged to consult classical textbooks like [27] in Chap. 6 or [48] in Chap. 3. Nevertheless much work has to be added. It is interesting to notice that this example supports Helmholtz' law that vorticity is convected by the outer flow.

As we have only $\omega_\phi = A \cdot r$ and

$$u_r = -\frac{1}{r}\frac{\partial \Psi}{\partial z}, \tag{A.85}$$

$$u_z = \frac{1}{r}\frac{\partial \Psi}{\partial r}, \tag{A.86}$$

we may integrate Eq. (6.66) via

$$H_r = \int -u_z \omega_\phi dr \text{ and} \tag{A.87}$$

$$H_z = \int u_r \omega_\phi dz. \tag{A.88}$$

6.5 Solution for problem: FORTRAN source code of vortex patch.

```fortran
c-----------------------------------------------+
c       (c) Schaffarczyk, UAS Kiel, Germany     +
c-----------------------------------------------+
c
c       first version 2006
c       vortex patch method for h-darrieus engines
c       after strickland et al trans asme 1979 79 wa/fe6 pp 1 ff
c
c       (c) march 2010
c       all data real*8 (= double precosion)
c           to improver numerical accuracy
c
c       nbl   = number of blades > 0
c       nvv   = number of vortex patches
c       nrevol = number or revolutions
c
c       nb l number of blades
c
        parameter (nbl   =     3)
        parameter (nvv   = 10000)
c
c       number of revolutions
c
        parameter (nrevol=     7)
c
        integer nx, ny
c
        real*8 circblnew(nbl), circblold(nbl)
        real*8 pb(nbl), blcoor(nbl,2)
        real*8 blvelu(nbl,3), blvind(nbl,3), vblind(nbl,3)
c
        real*8 vortgam(nbl*(nvv+1)), vortloc(nbl*(nvv+1),2)
        real*8 vortvel(nbl*(nvv+1),2), vortveloid(nbl*(nvv+1),2)
c
        real*8 dr(2), vvind(2), wind(2), rv(2), vec(2), tv(2)
        real*8 t, tx, ty, deg, nox, noy, pi, stp, dt
        real*8 aoa, cl, cd, vx, vy, v, Po, aw, uin
        real*8 ft, ftb, ct, fn, fnb, cn, dsq
c
        real*8 lam, sig, Liftf, Dragf, om, phi, dphit, gasch
        real*8 ak, gk, aku, gku, d2, istp, vovel
c
c       ***** debug flags ***************
c
        logical freewake, back onblade, wilson
        freewake    = .true.
        backonblade = .true.
        wilson      = .false.
c
c       ***** debug
c
c       set up initial position of blades
c
        pi  = 4.d0*datan(1.d0)
        deg = 1.8d2/pi
c
        dphibi = 2.d0*pi/nbl
c
c       H = 1 don't change
c
        H   = 1.0d0
```

```fortran
c---------------------------------------------------
c
c ************* input DATA ********************
c
c       r = radius of rotor
c       uin = inflow velocities
c       units are metric
c
        R    = 1.0d0
        uin  = 1.0d0
c
c       sig = solidity = B chord/Radius
c       lam(bda) = tip speed ratio
c
        sig  = 0.30d0
        lam  = 4.0d0
c
c *************************************************
c
        ch   = R*sig/nbl
        om   = lam*uin/R
c
        wind(1) = uin
        wind(2) = 0.0d0
        rho   = 1.225d0
        stp   = 1.500*rho*uin**2
        area  = 2.d0*R*H
c
c       time for one revolution
c
        t    = 2.d0*pi/om
c
c       tanf = time from which on transients are gone
c       here = assume 3 revolutions
c
        tanf = (nrevol-3.d0)*t
        tend = nrevol*t
c
        tint = tend-tanf
        tfre = nrevol*t
c
c       4 deg steps -> 90
c
        dt   = t/90.d0
        ntime = tend/dt
        write(*,*) '****************'
        write(*,'(a20,2f6.2,e10.3)') 'tanf,tend,dt',tanf,tend,dt
        write(*,*) '****************'
c
        dphit = om*dt
c
c       actual number of vortexpatches
c
        nvort = 0
        nt  = 0
c
        do i=1,nbl*(ntime+1)
          vortgam(i)=0.0
          do k=1,2
            vortloc(i,k)=0.0
          end do
        end do
```

```
c        shedded circulation = 0
c
         do i=1,nbl
            circbiold(i) = 0.0
            circbinew(i) = 0.0
         end do
c
c-----------------------------------------------------------------
c        time stepping t = n*dt, n > 0
c-----------------------------------------------------------------
c
         Pbar   = 0.0d0
         Fxbar  = 0.0d0
         Fybar  = 0.0d0
c
         write(11,208)'Phi(1)','aoau','aoa','a','cN','cX','cT','cP','10 cl'
c
c        time step loop
c
         write(*,*)'ntime =',ntime
         do while (nt.le.ntime)
c
            if(nt/100*100.eq.nt)write(*,*)'nt =',nt
c
            nt = nt+1
            tact = nt*dt
c
            do i=1,nbl*(ntime+1)
               do k=1,2
                  vortvel(i,k)    = 0.0
                  vortvelold(i,k) = 0.0
               end do
            end do
c
c           new azimuth of blade(s)
c
            do i=1,nbl
               pb(i) = pb(i)+dphit
            end do
c
c**** debug output on display ************************************
c
            if(nt/10*10.eq.nt) then
               phiout=deg*pb(1)
               write(*,201) t=',tact,' phi: ',phiout,' nv:',nvort
            endif
c****************************************************************
c
c           new cartesian coordinates
c
            do i=1,nbl
               blcoor(i,1) = r*dcos(pb(i))
               blcoor(i,2) = r*dsin(pb(i))
            end do
c
c           new driven velocities (x = 1, y = 2, absolute value = 3) relative to b
lades
```

```
c
         write(*,'(a12,i8)')'b      =',nbl
         write(*,'(a12,f8.3)')'la*sig=',lam*sig
         write(*,'(a12,f8.3)')'la    =',lam

         open(unit=10,file='out',form='formatted',status='unknown')
         open(unit=11,file='for1',form='formatted',status='unknown')
         open(unit=12,file='vec1',form='formatted',status='unknown')
         open(unit=13,file='cir',form='formatted',status='unknown')
c
c------------------------------------------------------------------
c        initial values t = 0
c------------------------------------------------------------------
c
c        azimuthal angle(s) of blade(s)
c        phi = 0 along x-axis
c
         do i=1,nbl
            pb(i)= (i-1)*dphibl
         end do
c
c        cartesian coord (x = 1, y = 2)
c
         do i=1,nbl
            blcoor(i,1) = r*dcos(pb(i))
            blcoor(i,2) = r*dsin(pb(i))
         end do
c
c        wind velocites relative to blades (x = 1, y = 2, absolute value = 3)
c
         do i=1,nbl
            blvelu(i,1) = uin +om*r*dsin(pb(i))
            blvelu(i,2) =     -om*r*dcos(pb(i))
            blvelu(i,3) = dsqrt(blvelu(i,1)**2+blvelu(i,2)**2)
         end do
c
c        angle of attack -> lift -> circulation on blade -> forces
c
         do i=1,nbl
            phi = pb(i)
            vx  = blvelu(i,1)
            vy  = blvelu(i,2)
            v   = blvelu(i,3)

            tx  =  -dsin(phi)
            ty  =   dcos(phi)
            nox =   dcos(phi)
            noy =   dsin(phi)

            gk  =  nox*vx+noy*vy
            ak  = -(tx*vx+ty*vy)
            aoa =  datan2(gk,ak)
c
            call lift(aoa,cl)
            circbiold(i) = 0.5*cl*ch*blvelu(i,3)
         end do
```

```fortran
      aw = 0.0d0
      if (wilson)then
        aw = 0.5d0*sig*lam*dabs(dcos(phi))
        if(aw.gt..5d0)aw=.5d0
        a = aw
      endif

c
      do i=1,nbl
        bivelu(i,1) = uin +om*r*dsin(pb(i))
        bivelu(i,2) = -om*r*dcos(pb(i))
        bivelu(i,3) = dsqrt(bivelu(i,1)**2+bivelu(i,2)**2)
      end do

c
      do i=1,nbl
        bivelu(i,1) =  bivelu(i,1)-aw*uin
        bivelu(i,2) =  bivelu(i,2)
      end do

c
c     correct blade velocity from OLD vortex system (n-1)*dt
c     - if there is any -
c
c
      if (backonblade)then
c
        do i=1,nbl
        do k=1,2
          vblind(i,k) = 0.0d0
        end do
        end do
c
        if(invort.ge.nbl)then
        do i=1,nbl
        do j=1,nvort
          dsq = 0.0d0
        do k=1,2
          dr(k) = bicoor(i,k)-vortloc(j,k)
          dsq   = dsq+dr(k)**2
        end do
          core = 0.5d0*ch
          if (dsqrt(dsq).gt.core)then
          vblind(i,1)=vblind(i,1)-vortgam(j)*dr(2)/(2.d0*pi*dsq)
          vblind(i,2)=vblind(i,2)+vortgam(j)*dr(1)/(2.d0*pi*dsq)
          end if
        end do
        end do
        end if
        endif (backonblade)
c
        do i=1,nbl
          vblind(i,3) = dsqrt(vblind(i,1)**2+vblind(i,2)**2)
        end do
        endif
c
        do i=1,nbl
        do k=1,2
          bivel(i,k) = bivelu(i,k) + vblind(i,k)
        end do
        end do
c
```

```fortran
c     calculate absolute value of blade velocity
c
      do i=1,nbl
        bivel(i,3) = dsqrt((bivel(i,1))**2+(bivel(i,2))**2)
      end do

c
c-------------
c     calculate force-data and new shedded vorticity
c
c     "U" meaning uncorrected without vortex induces velocities
c
c-------------
c
      Fn = 0.0d0
      Ft = 0.0d0
      Fx = 0.0d0
      Fy = 0.0d0
      Po = 0.0d0
      a  = 0.0d0
c
c     BLADE LOOP
cbbbbbbbbbbbbbbbbbbbbbbbbbbbbbbbbbbbbbbbbbbbbbbbbbbbbbbbbbbbbbbbbbbbbbbbbbb
      do i=1,nbl
        phi = pb(i)
        vx = bivel(i,1)
        vy = bivel(i,2)
        v  = bivel(i,3)
c
        vxu= bivelu(i,1)
        vyu= bivelu(i,2)
        vu = bivelu(i,3)
        ab = vblind(i,1)/uin
c++++++++++++++++++++++++++++++++++++++++++++++++++++++++++++++++++++++++++
        nox =  dcos(phi)
        noy =  dsin(phi)
c
        tx = -dsin(phi)
        ty =  dcos(phi)
c
c     opposite leg (in german: Gegenkathete) = project to  normal(radial
) direction)
c
        gk = nox*vx+noy*vy
c
c     !!! think carefully of signs !!!!
c
        ak  = -(tx*vx+ty*vy)
c
        aoa = datan2(gk,ak)
c
        gku = nox*vxu+noy*vyu
        aku = -(tx*vxu+ty*vyu)
        aoau= datan2(gku,aku)
c
c++++++++++++++++++++++++++++++++++++++++++++++++++++++++++++++++++++++++++
c************ debug aoa = angel of attack ******************
c
c            nach gasch 3. aufl. pp 350
```

```fortran
c
      if (wilson)a = aw
c
      write(11,206)deg*pb(i),r1,r2,a,cn,cx,ct,cpp,l.di*cl
c
c------------------------------------------------------------
c    CREATE shedded vorticity from each blade and its initial location
c------------------------------------------------------------
c
      do i = 1, nbl
c
c    here we have creation of a new vortex
c
      nvort = nvort + 1
c
c    sheded vorticity is -dGamma
c
      vortgam(nvort)= - circblold(i) + circblnew(i)
      circblold(i) = circblnew(i)
      tv(1) = -0.5d0*ch*dsin(phi)
      tv(2) = 0.5d0*ch*dcos(phi)
      do k=1,2
         vortloc(nvort,k)=blcoor(i,k)-tv(k)
      end do
      write(*,203)'CR gam ',vortgam(nvort),' binr: ',i
      end do
c
c    set initial velocity of new (last nbl) vortices to induced blade velo
city
c
      nvold = nvort-nbl
c
      do i=1,nbl
         newv = nvold+i
         do kc=1,2
            vortvel(newv,kc) = vblind(i,kc) +wind(kc)
         end do
         vovel = dsqrt(vortvel(newv,1)**2+vortvel(newv,2)**2)
         write(13,204)newv,vortgam(newv),vovel
      end do
c
c------------------------------------------------------------
c
c    1) store old values
c
      do i=1,nvold
         do k = 1,2
            vortvelold(i,k)=vortvel(i,k)
            vortvel(i,k)   = 0.0d0
         end do
      end do
      do i=nvold+1,nbl*(ntime+1)
         do k = 1,2
            vortvelold(i,k)=0.0d0
         end do
      end do
c
c    2) now calculate influence of bound vorticity
c
      do jref=1,nvort
```

```fortran
c
c
c
c
      aoa  = dasin(uin*dcos(phi)/(om*r))
      aoau = dasin(uin*dcos(phi)/(om*r))
c
c*************************************************
c
      call lift(aoa,cl)
      call drag(aoa,cd)
      lstp = 0.5d0*rho*v**2
      Liftf    =        cl*lstp*ch*h
      Dragf    =        cd*lstp*ch*h
c-----------------------------------------------------
c    new vorticity around blade
c
      circblnew(i) = cl*ch*v/2.d0
c
c-----------------------------------------------------
c    force data
c
      Ftb = Liftf*dsin(aoa)-Dragf*dcos(aoa)
      Fnb = Liftf*dcos(aoa)+Dragf*dsin(aoa)
c
      Fxb = Fnb*dcos(phi)-Ftb*dsin(phi)
      Fyb = Fnb*dsin(phi)+Ftb*dcos(phi)
c
      write(*,'(a15,3f8.2)')'ftb fnb fxb',ftb,fnb,fxb
c
      Ft = Ft + Ftb
      Fn = Fn + Fnb
      Fx = Fx + Fxb
      Fy = Fy + Fyb
c
      a   = a + ab/nbl
c
      Po  = Po + Ftb*R*om
      end do
c    END BLADE LOOP
c
cbbbbbbbbbbbbbbbbbbbbbbbbbbbbbbbbbbbbbbbbbbbbbbbbbbbbbbbbbbbbbb
c
      write(*,'(a15,3f8.2)')'ft fn fx',ft,ft,fn,fx
c
      if (tact.ge.tanf)then
         Pbar  = Pbar + Po*dt
         Fxbar = Fxbar + Fx*dt
         Fybar = Fybar + Fy*dt
      endif
c
      cn = Fn/(stp*area)
      cx = Fx/(stp*area)
      cy = Fy/(stp*area)
      ct = Ft/(stp*area)
      cpp = Po/(uin*stp*area)
      r1 = deg*aoau
      r2 = deg*aoa
c
      write(*,'(a15,3f8.2)')'ct cn cx',ct,cn,cx
c
c    azimuthal printout
```

| Jul 26, 13 16:15 | VortexPatchDarrieus.f | Seite 10/12 |

```
              vvind(2) = vindu*dr(1)/(2.*pi*d2)
              do k=1,2
                vortvel(jref,k)=vortvel(jref,k) + vvind(k)
              end do
 100        continue
          end do
        end if
c
c       jref loop
c
        end do
c
c------------------------------------------------------------------
c       adjust all locations with induced velocities + wind (eg  (8)
c
c       use "open" differencing schema of strickland loc. cit. eq (8)
c       use "freezed" wake if t > tfre
c
        if (tact.ge.tfre)then
          w1 =  0.0d0
          w2 =  0.0d0
        else
          w1 =  1.5d0
          w2 = -0.5d0
        end if
c
c       konvektions geschw der wirbel bestimmen !!
c
        do i=1,nvort
          do k=1,2
            vex = w1*vortvel(i,k)+w2*vortvelold(i,k)+wind(k)
            vortloc(i,k) = vortloc(i,k)+vex*dt
          end do
        end do
c
c------------------------------------------------------------------
c       now print out vortex locations
c
c
        if(nt.eq.ntime)then
          do kv=1,nbl
            do i=kv,nvort,nbl
              write(10,200)i,(vortloc(i,ko), ko=1,2),vortgam(i)
            end do
            write(10,200)
          end do
        end if
c
c------------------------------------------------------------------
c       END time stepping LOOP
c
        end do
c
c       average over integration time
c
        cx   = Fxbar/(    stp*area*tint)
        cy   = Fybar/(    stp*area*tint)
        cp   = Pbar/(uin*stp*area*tint)
```

| Jul 26, 13 16:15 | VortexPatchDarrieus.f | Seite 9/12 |

```
c
c       a) influence of bound vortices on vortex patches
c
        if(jref.it.nvold)then
          do ib =1,nbl
            d2 = 0.0d0
            do k=1,2
              dr(k) = vortloc(jref,k)-blcoor(ib,k)
              d2 = d2+dr(k)**2
            end do
c
c           cut-off procedure in case vortices are too close
c
            core  = ch
            vindu = circblnew(ib)
            d     = dsqrt(d2)
            if(d.it.core)then
              d2 = core**2
              do k4=1,2
                dr(k4)=core*dr(k4)/d
              end do
            end if
c
c           use Biot's and Savart's law for "induction"
c
            vvind(1) = -vindu*dr(2)/(2.*pi*d2)
            vvind(2) =  vindu*dr(1)/(2.*pi*d2)
c
            do k=1,2
              vortvel(jref,k)=vortvel(jref,k) + vvind(k)
            end do
          end do
        end if
c
c       b) influence of other vortices
c
        if (freewake) then
c
          do indu=1,nvort
c
c           assume NO influence of vortex on itself
c
            if (indu.eq.jref)goto 100
            d2 = 0.0
            do k=1,2
              dr(k) = vortloc(jref,k)-vortloc(indu,k)
              d2 = d2+dr(k)**2
            end do
c
c           cut-off procedure in case vortices are too close
c
            core  = ch
            vindu = vortgam(indu)
            d     = dsqrt(d2)
            if(d.it.core)then
              d2 = core**2
              do kk=1,2
                dr(kk)=core*dr(kk)/d
              end do
            end if
c
            vvind(1) = -vindu*dr(2)/(2.*pi*d2)
```

```fortran
      write(*,211)'T avg:',lam sig',lam,sig
      write(*,207)'cx,cy',cx,cy
      write(*,207)'cm,ct',cn,ct
      write(*,207)'cp=',cp=,cp
c
c
c     print out velocity-vector field
c

      nx = 30
      ny = 50
      xa = -4.d0
      xe = 18.d0
      ya = -5.d0
      ye = 5.d0
      vscal = 0.6d0

c
      dx = (xe-xa)/nx
      dy = (ye-ya)/ny

c
      do ix = 1,nx
        rv(1) = xa+ix*dx
        do iy = 1,ny
          rv(2) = ya+iy*dy
          do k=1,2
            vvind(k)=0.0
          end do
          do j=1,nvort
            dsq = 0.0
            do k=1,2
              dr(k) = rv(k)-vortloc(j,k)
              dsq = dsq+dr(k)**2
            end do
            vvind(1) = vvind(1)-vortgam(j)*dr(2)/(2.*pi*dsq)
            vvind(2) = vvind(2)+vortgam(j)*dr(1)/(2.*pi*dsq)
          end do

c     hier einfluss der nbl bound vortices auf das v-feld

          do ib =1,nbl
            d2 = 0.0
            do k=1,2
              dr(k)= rv(k)-blcoor(ib,k)
              d2 = d2+dr(k)**2
            end do

c     cut-off procedure in case vortices are too close

            core= ch
            vindu = circblnew(ib)
            vindu = 0.0
            d  = sqrt(d2)
            if(d.lt.core)then
              d2 = core**2
              do kk=1,2
                dr(kk)=core*dr(kk)/d
              end do
            end if

c
            vvind(1) = vvind(1)-vindu*dr(2)/(2.*pi*d2)
            vvind(2) = vvind(2)+vindu*dr(1)/(2.*pi*d2)
c
          end do

          do k=1,2
            vec(k) = wind(k)+vvind(k)
            vec(k) = vscal*vec(k)
          end do
          write(12,209)(rv(k), k=1,2),(vec(k),k=1,2)
        end do
      end do

c
      close(unit=10)
      close(unit=11)
      close(unit=12)
      close(unit=13)
c
c
200   format(i5,3f12.6)
201   format(a6,f8.2,a6,f6.0,a6,i5)
202   format(a10,3f12.2)
203   format(a10,f12.6,a10,i2)
204   format(i5,2f12.6)
205   format(a10,2f12.6)
206   format(3f8.2,6f12.4)
207   format(a5,2f10.4)
208   format(3f8.6,i2)
209   format(4f12.6)
211   format(a16,2f6.2)
c
c ----------------------------- end main program -----------------------------
c
      end
c
c -----------------------------------------------------------
c
      subroutine lift(al,cl)
      real*8 al, cl, clmax, zwpi

      zwpi = 6.2832d0
      clmax = 1.5

      cl = zwpi*al
      if(cl.gt. clmax) cl =  clmax
      if(cl.lt.-clmax) cl = -clmax

      return
      end
c
c -----------------------------------------------------------
c
      subroutine drag(al,cd)
      real*8 al, cd

      cd = 0.0d0

      return
      end
```

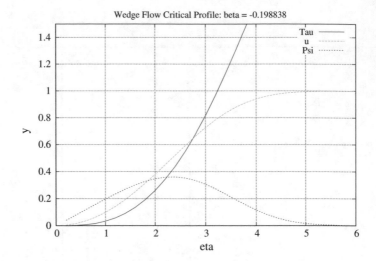

Fig. A.27   Graphical display of solution to Problem 7.3

A.7   Solutions for Problems of Sect. 7.10

7.1 Logarithmic profile and k-ε model [26] in Chap. 7:
   Introduce scaling variables:

$$0 - c_\mu \frac{u^{\star 4}}{c_\mu} \frac{\chi y}{u^{\star 3}} \frac{u^{\star 2}}{\chi^2 y^2} + \frac{u^{\star 3}}{\chi y} = 0 \,, \tag{A.89}$$

$$-\frac{c_\varepsilon}{c_\mu} \frac{u^{\star 4}}{y^2} - c_1 \frac{u^{\star 4}}{\sqrt{c_\mu}} \chi^2 y^2 + c_2 \sqrt{c_\mu} \frac{u^{\star 4}}{\chi^2 y^2} \,. \tag{A.90}$$

7.2 Decaying turbulence and k-ε model [26] in Chap. 7:
Our solution—as in Problem 7.1—is from [26] in Chap. 7: Let $\tau := 1 + \lambda t$. Inserting
in Eqs. 7.82 and 7.83 gives

$$m = n + 1 = -n + 2m - 1 \,, \tag{A.91}$$

$$c_2 = n + \frac{1}{n} \,. \tag{A.92}$$

With $n = 1.3$ given we get $c_2 = 2.06$.

7.3 Critical Falkner–SKan (wedge) flow:
Figure A.27 shows the critical values just before separation $\beta = 0.19884$ [64] in
Chap. 3.

7.4 XFoil for DU-W-300 (Figs. A.28, A.29, and A.30):
Measurement at K"ølner KK (2002/2003) (Figs. A.31, A.32, A.33, A.34, A.35, and
A.36).

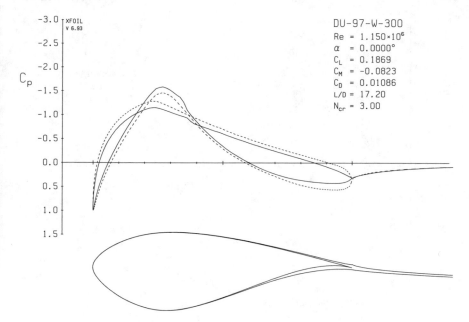

Fig. A.28 XFoil results for DU-W-300, Re = 1.5 M, AOA 0 deg

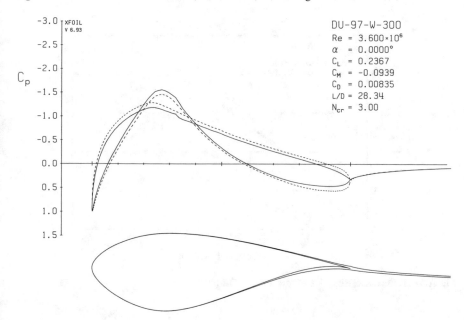

Fig. A.29 XFoil results for DU-W-300, Re = 3.6 M, AOA 0 deg

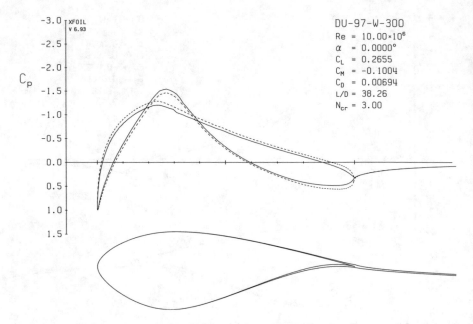

Fig. A.30 XFoil results for DU-W-300, Re = 10 M, AOA 0 deg

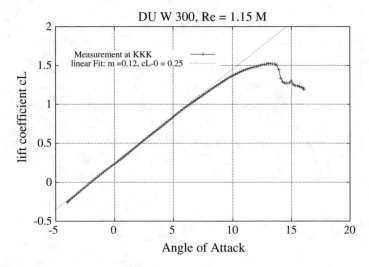

Fig. A.31 Kölner KK, Re = 1.2 M: c_L versus AOA (α)

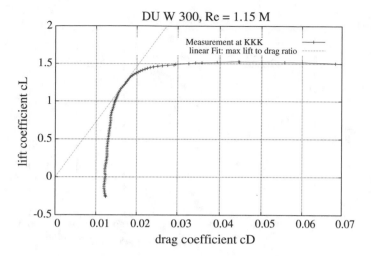

Fig. A.32 Kölner KK, Re = 1.2 M: c_L versus c_D

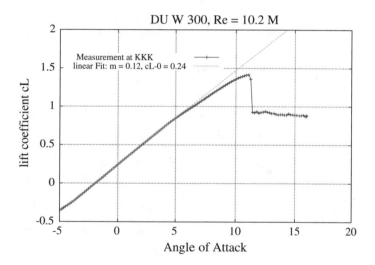

Fig. A.33 Kölner KK, Re = 3.6: c_L versus angle of attack (α)

7.5 CFD code (FLOWER) with e^N transition module (Fig. A.37).

A.8 Solutions for Problems of Sect. 8.9

8.1 Estimation of lift by pressure integration

$$c_{p,i} := \frac{c_{p,i} + c_{p,i+1}}{2},$$
(A.93)

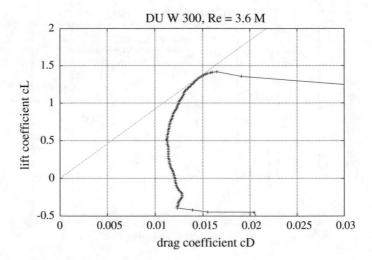

Fig. A.34 Kölner KK, Re = 3.6 M: c_L versus c_D

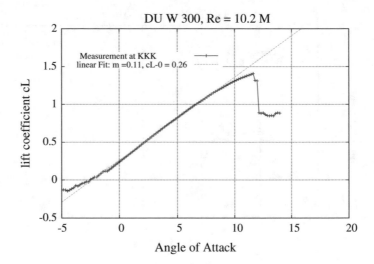

Fig. A.35 Kölner KK, Re = 10 M: c_L versus angle of attack (α)

$$dx_i := x_{i+1} - x_i, \tag{A.94}$$

$$dy_i := y_{i+1} - y_i, \tag{A.95}$$

$$c_N = \sum_{i=1}^{N} c_{p,i} \cdot dx_i, \tag{A.96}$$

$$c_T = \sum_{i=1}^{N} c_{p,i} \cdot dy_i. \tag{A.97}$$

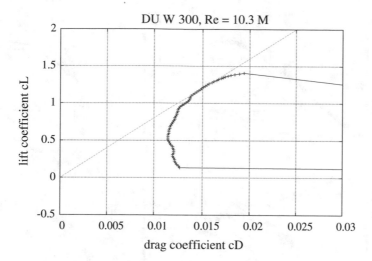

Fig. A.36 Kölner KK, Re $= 10$ M: c_L versus c_D

With AOA α known, lift may be calculated via

$$c_L = c_N \cdot cos(\alpha) - c_T \cdot sin(\alpha) . \tag{A.98}$$

The results should be $c_L(\alpha = 0) = 0.214$ and $c_L(\alpha = 8) = 1.019$, respectively.

8.2 Estimation of a power curve (Fig. A.38).

8.3 Lift-to-drag ratio for measured data (Fig. A.39 and Table A.7).

A.9 Solutions for Problems of Sect. 9.6

9.1 Wind Vehicle Part 2: Blade design.

Summary from Problem 5.5

Maximum power extraction is not meaningful for a wind-driven vehicle. Instead we chose $v_{car}/v_{wind} \rightarrow max$. In addition we use general BEM. As usual we assume all ring segments to be independent. No radial flow is present and lift is assumed to be the only force acting. Comparing with Fig. A.40, we have

$$dP = \omega r \cdot dL \cdot cos(\varphi) \tag{A.99}$$

with

$$dT = dL \cdot sin(\varphi). \tag{A.100}$$

Now lift is

$$dL = \frac{\rho}{2}w^2 \cdot c_L \cdot c(r) \cdot dr \tag{A.101}$$

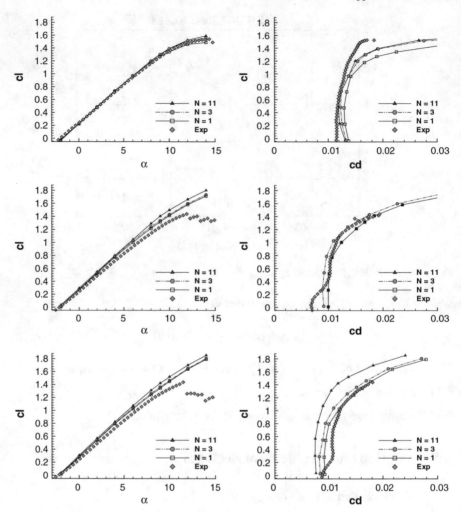

Fig. A.37 DU300-mod: $c_l(\alpha)$ and $c_l(c_d)$ for comparison with measurements at Re $1.2 \cdot 10^6$, $5.8 \cdot 10^6$, and $10.2 \cdot 10^6$

and the total inflow velocity:

$$w^2 = ((1 - a) \cdot v_{wind})^2 + (\omega(1 + a'))^2 . \tag{A.102}$$

a and a' again are the usual induction factors [44], in Chap. 5.

Two ways for optimization have been described:

- Optimize local fluid angle φ, and choose chord c(r) as usual [20, 21] in Chap. 9;
- Optimize c(r) and choose AOA so that the profile works at maximum c_L/c_D [9], in Chap. 5.

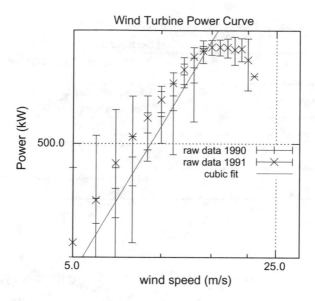

Fig. A.38 Double logarithmic plot

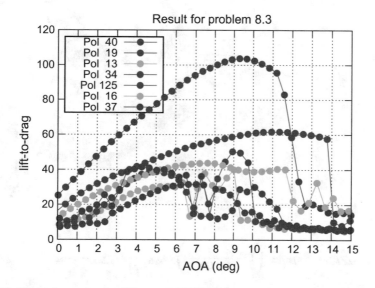

Fig. A.39 Lift-to-drag ratio of well-known DU97-W-300-mod

Stuttgart [20, 21] in Chap. 9 gives numbers for comparison. Their lightly loaded ($a = 0.2$) (2008) rotor had $\lambda = 5$, resulting in $c_P = 0.24$ and $c_T = 0.23$. At $v_{wind} = 8\,\mathrm{m/s}$ the car's speed was estimated to $v_{car}^{max} = 5m/s$ using a power of $P_{Max} = 250\,\mathrm{W}$ only.

The Danish design [44] in Chap. 5 may be summarized in equation

Table A.7 Explanations of polars, Ma = 0.2 , Re = 2.2 M

Number	Description
13	Severe roughness (Karborundum 60) around nose
16	Tripping wire at x/c = 0.15
19	Tripping wire at x/c = 0.30
34	As 13 but with VGs at x/c = 0.35
37	As 13 but with 2% GF
40	Clean
125	Less sever roughness (Karborundum 120) around nose

Fig. A.40 Blade segments

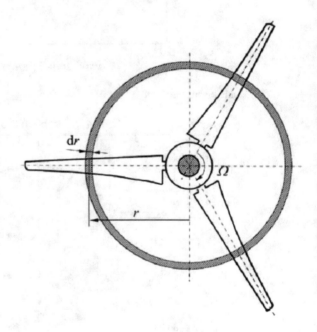

$$C_{PROPF,loc} = \eta_P \eta_T \left(1 + \frac{1}{V/V_{wind}}\right) C_{P,loc} - C_{T,loc} \to \max. \qquad (A.103)$$

Blade Design Method

We use de Vries [44] in Chap. 5 and start with

$$C_T = \frac{8}{\lambda^2} \int_0^\lambda (1 - aF) aF \left(1 + \frac{tan(\varphi)}{L2D}\right) x \, dx, \qquad (A.104)$$

$$C_P = \frac{8}{\lambda^2} \int_0^\lambda (1 - aF) aF \left(tan(\varphi) - \frac{1}{L2D}\right) x^2 \, dx. \qquad (A.105)$$

Table A.8 Sample results from race 2019 at Den Helder, The Netherlands

Fahrzeug	Rang	Endurance %	Drag race	Fastest %	Innovation
Chinook	1	105.5	1	114.8	2
BTT	2	59.2	2	78.4	1
Wind turbine racer	3	72.9	3	103.3	8
NeoVento-Spirit of Amsterdam	4	27.3	4	47.7	4
DTU E	5	20.9	–	40.4	7
Stormy Sussex	6	–	–	–	2
REK Yildiz	7	–	–	–	5
Fibershark–Alkmaar	8	–	–	-	6

Here L2D is the usual lift-to-drag ratio and $x = \lambda \rho$ a local tip–speed ratio. For tip-correction we use Eq. 5.59 from Chap. 5:

$$F = \frac{2}{\pi} \arccos \left(\exp \left(-B/2 \ (1 - \rho) / (\rho \sin (\varphi) \right) \right) \tag{A.106}$$

introducing $\rho = r/R$ as the normalized distance to tip.
   Optimization of Eq. (A.103)

$$C_{PROPF,loc} \rightarrow max \tag{A.107}$$

is performed numerically using a binary-search algorithm. The (FORTRAN) source code is shown from Sect. A.9 on.

Remark Results from this optimization are most easily attained by introducing a dimensionless drag coefficient of the car as shown in Problem 5.5:

$$K := C_D \cdot \frac{A_V}{A_R} . \tag{A.108}$$

C_D is the bare drag coefficient of the vehicle without a rotor $D := c_D \cdot \rho/2v^2 \cdot A_V$, A_V and A_R are projected areas of the rotor and the vehicle, see Fig. A.17.

   Inclusion of a constant rolling resistance seems to be problematic. After solving the equations blade geometry (chord- and twist-distribution) may be deduced, see Fig. A.42. For design we used $B = 3$, $\lambda = 5.5$, design lift $C_L^{des} = 1, 0$, and L2D $= 80$. Diameter of the rotor is 1800 mm and drivetrain efficiency is assumed to be $\eta_{Drivetrain} = 0.7$. An interesting question is whether the same results (decrease of loading) may be achieved by simply pitching an ordinary wind turbine blade

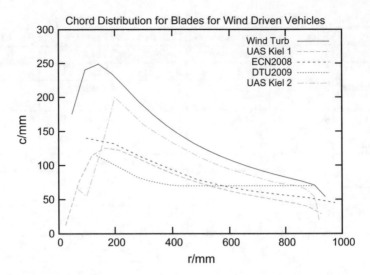

Fig. A.41 Chord distribution of different designs

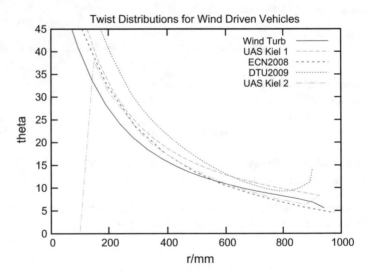

Fig. A.42 Twist distribution of different designs

(Wind Dynamics—3WTR for example). Figure A.19 compares the results from our method with those from Sørensen's [29] in Chap. 9 approach. It has to be noted that our values should be generally smaller because more effects of losses (C_D tip-correction) are included. Figure A.19 shows further which limit K-values may be achieved if unrealistically high c_P are excluded. A further insight is that the reported values from [21] in Chap. 9 may only be possible if $K \leq 0.005$ which is extremely small (Figs. A.41 and A.41).

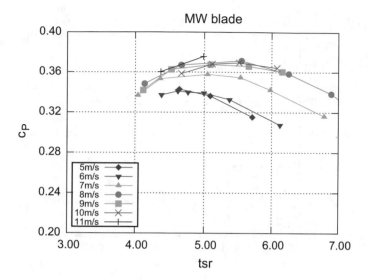

Fig. A.43 c_P for the multi-wind blades

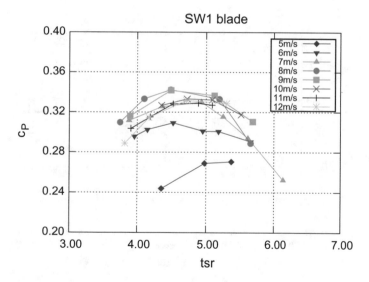

Fig. A.44 c_P for the SW-1 blades

Wind Tunnel Entries

Here we summarize results from wind tunnel entries with a wind turbine blade (MW) and an own design (SW-1) (Figs. A.43, A.44, A.45, A.46, A.47, and A.48) .

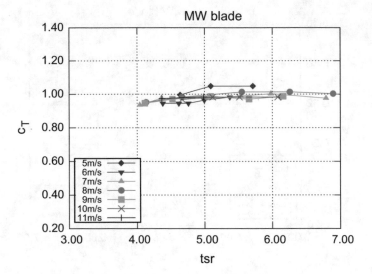

Fig. A.45 c_T for the multi-wind blades

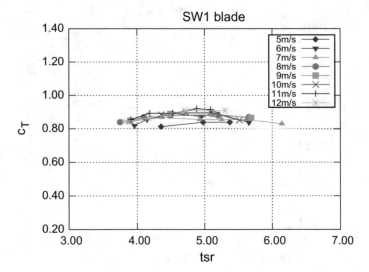

Fig. A.46 c_T for the SW-1 blades

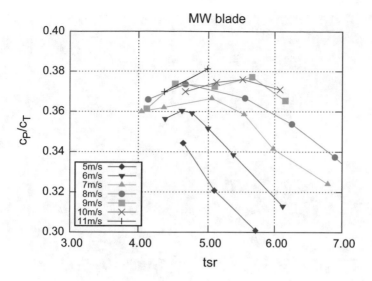

Fig. A.47 c_P/c_T of the multi-wind blades

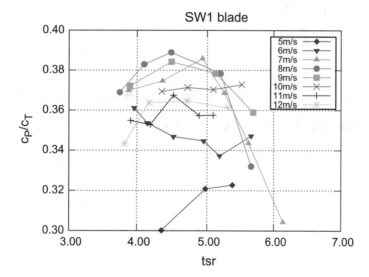

Fig. A.48 c_P/c_T of the SW-1 blades

Fig. A.49 Our car of 2019, 2nd rank

Wind Car Baltic Thunder

Finally we give an impression we see in Fig. A.49 of our car from 2019. It received second rank in Table A.8. Figure A.18 already showed a comparison of all results so far.

FORTRAN source: Wind Vehicle Design Code

```fortran
c
c-------------------------------------------------------
c-------------------------------------------------------
c
      PROGRAMM opt  (C)  A.-P. SCHAFFARCZYK UAS Kiel
c     march 1998, nov 2004, oct 2009
c
c     for CHINOOK, 23. august 2011
c
c     basic equation come from DEVRIES AGARDOgraph 243 (1979), Kap 4.4 ff
c     and
c     Mac Gaunaa et.al. proc. ewec 2009, Equation (36) - attached
c
c     some commends are in GERMAN - remember my discussion afer some beers ....
c
c
      COMMON PI, B, LAMBDA, GZ, vratio, etadr
c
      CHARACTER*100 NAMEOUT1, NAMEOUT2, NAMEOUT3
      CHARACTER*100 NAMEIN
      CHARACTER*6   s1, s2
      CHARACTER*6   steuer
c
      REAL PI, LAMBDA, la, B, GZ, etadr
c
C--- NUMERIC constants
c
      PI       = 4. * ATAN(1.)
      EPS      = 1.E-07
      EPSG     = 1.E-05
      MAXI     = 200
      MAXJ     = 100
      thetamax = pi/2.
c
c********* start parameter ******************************
c*******************************************************
c
c--- to change: re-compile
c
c     number of B(lades), Lambda = Omega R_tip = TSR , G(leit)Z(ahl) = L2D ratio
c     Rtip = Tip-Radius
c     Cl-des desing lift
c
      B      = 3.0
      LAMBDA = 5.5
      GZ     = 70.0
      Rtip   = 0.95
c
c     design lift
c
      Cldes  = 1.00
c
c     aoa (Cl-des)
c
      aoa    = 6
```

```fortran
c     vratio = desing ratio;
c     etadr = drive-train efficieny
c
      vratio =   0.5
      etadr  =   0.7
cc***************** end parameter *********************
cc
cc
c !!!!!!!!!!!!!!! DON'T CHANGE from here !!!!!!!!!!!!!!!!!!
c
      write(*,'(2(a10,0f6.3))')'vra=',vratio,'etu=',etadr
c
      drag   = 1./gz
c
      intla  = int(10*lambda)
      i1 = intla/10
      i2 = intla-10*i1
      s1 = ACHAR(48+i1)
      s2 = achar(48+i2)
c
      IOOUT21 = 21
      OPEN(UNIT=IOOUT21,FILE='list',FORM='FORMATTED',STATUS='UNKNOWN')
      WRITE(21,'(5a12)')'B','LAMBDA','GZ','Rtip','Design CL'
      WRITE(21,'(5f12.4)') B, LAMBDA, GZ, Rtip,CLdes
c
      Write(21,*)
c
      NAMEOUT1 = 'BTdes.' // s1 //s2
c
      IOOUT1 = 12
      OPEN(UNIT=IOOUT1,FILE=NAMEOUT1,FORM='FORMATTED',STATUS='UNKNOWN')
      WRITE(IOOUT1,'(7a12)')
     + 'x=/Rtip','xx=*la','c Cl/Rtip','r/mm','c/mm','Phi/Deg','F'
      WRITE(*,'(7a12)')
     + 'x=/Rtip','xx=*la','c Cl/Rtip','r/mm','c/mm','Phi/Deg','F'
c
      NAMEOUT2 = 'INDUCTION_FACTORS'
c
      IOOUT2 = 13
      OPEN(UNIT=IOOUT2,FILE=NAMEOUT2,FORM='FORMATTED',STATUS='UNKNOWN')
      WRITE(IOOUT2,'(5i12)')'X','X ax','qphi','dcp','A phi'
c
      NAMEOUT3 = 'steuer.dat'
c
      IOOUT3 = 14
      OPEN(UNIT=IOOUT3,FILE=NAMEOUT3,FORM='FORMATTED',STATUS='UNKNOWN')
      WRITE(IOOUT3,'(5a10,a5)')'fn','z','pitch','chord','aec','ipr'
c
      aec    = 0.25
      ipr    = 0
      steuer = 'p1.dat'
c-
c
      cptot = 0.0
      cttot = 0.0
c
c     blade loop
```

```fortran
      nx = 20
      dx = 1./nx
c
      DO IX=0,nx-1
c
      x = r/rtip
      xx = x * lambda
c
      X  = (0.5+ix)*dx
      XX = LAMBDA*X
c
c     start from AD
c---  AUSGANGSWERTE NACH ACT DISK SCHAETZEN
c
c     deVries equation (4.4.18)
c
      THETANULL = .5*ATAN(1./XX)
      SIGMANULL = 4.0*(1.-COS(THETANULL))
c
      CCLACT   = 2.*PI*X*SIGMANULL/B
      write(*,*)'x sigmanull ',x,sigmanull
      DCPALIT  = DELITACP(X,SIGMANULL,THETANULL)
      DCTALIT  = DELITACT(X,SIGMANULL,THETANULL)
c
      WRITE(*,*)' from ACTUATOR DISK-THEORY:'
      WRITE(*,*)
      WRITE(*,*)'THETANULL(GRAD): ',180.*THETANULL/PI
      WRITE(*,*)'CCL             ',CCLACT
      WRITE(*,*)'DCPALIT:        ',DCPALIT
      WRITE(*,*)'DCTALIT:        ',DCTALIT
      WRITE(*,*)
      WRITE(*,*)
c
      DELTATHETA = .05*THETANULL
c
      WRITE(*,*)'DELTATHETA(GRAD):',180.*DELTATHETA/PI
c
c---  outer (theta) ITERATION
c     maximize (local !) NOT cP but :
c
      cP-car = eta*(1+ v-wind/v-car)*cP  - cT
c
      WRITE(*,101)'THETA          DCP          DF
C101  FORMAT(19X,A42)
c
      J = 0
c
 10   DO WHILE(.TRUE. .and.theta.lt.thetamax)
      J = J + 1
c
      IF (J.EQ.1)THEN
      THETA = THETANULL
      FALIT = DELITACPauto(X,SIGMANULL,THETANULL)
      ENDIF
c
c---  INNer (SIGMA) Iteration to
c     check if constraint G(SIG,THETA) = 0 is fulifilled
c
c     from wilson-lissaman (deVries Eq. 4.4.14)
c     may be some kind of orthogonality condition
```

```fortran
c
      I = 0
      DO WHILE(.TRUE.)
      I=I+1
      IF (I.EQ.1)THEN
      SIGMA = SIGMANULL
      ELSE
      SIGMA = SIGMA
 1    + - GRX(X,SIGMA,THETA)/DGDSIGMA(X,SIGMA,THETA)
      ENDIF
c
      GRXD = GRX(X,SIGMA,THETA)
c---
c     stop INNER ITERATION:
c
      IF(ABS(GRXD).LT.EPSG.OR.I.GT.MAXI)EXIT
      IF(I.GT.MAXI)WRITE(*,*)'WARNing:MAXIreached'
c
c---  END INNER ITERATION
c
      WRITE(*,*)I,' I,INN. ITER.: SIG,TH,GRX ',SIGMA,THETA,GRXD
c
      FNEU  = DELITACPauto(X,SIGMA,THETA)
      DF    = FNEU - FALIT
      DFEPS = ABS((FNEU - FALIT)/FNEU)
c
      WRITE(*,100)J,' AEUSS. ITER: THETA,DCP ',180.*THETA/PI,FNEU,DF
c100  FORMAT(I4,A14,3E14.5)
c
      IF (DF.GT.EPS)THEN
      THETA = THETA + DELTATHETA
      FALIT = FNEU
      ELSE
      DELTATHETA = -0.5*DELTATHETA
      THETA      = THETA + DELTATHETA
      FALIT      = FNEU
      ENDIF
c
      IF (DFEPS.LT.EPS.OR.J.GT.MAXJ)EXIT
      END DO
c---  END outer iteration
c
c
      WRITE(21,*)
      WRITE(21,*)
      WRITE(21,*)'--- ENDE DER ITERATION: I,J',I,J
      WRITE(21,*)'--- ENDE DER ITERATION: '
      WRITE(21,*)'--- ENDE DER ITERATION: '
c
      CCLOUT = CCL(X,SIGMA,THETA)
      AXOUT  = AAX(X,SIGMA,THETA)
      PHIOUT = APHI(X,SIGMA,THETA)
c
      WRITE(21,*)'CCL,THETA      ',CCLOUT,180.*THETA/PI
      WRITE(21,*)'AAX,APHI       ',AXOUT,PHIOUT
c
c     compute increment of  cP and  cT bestimmen
```

Gedruckt von CFD

Sep 05, 13 14:15	RA.f

```
      FUNCTION AAX (X,SIG,THETA)
C
C     COMMON PI,B,LAMBDA,GZ,vratio,etadr
C
C     REAL LAMBDA
C
      AAX = SORT(F(X,THETA)*F(X,THETA)+4.*A(SIG,THETA)*F(X,THETA)*
     1       (1.-F(X,THETA)))
      AAX = (2.*A(SIG,THETA)+F(X,THETA)-AAX)
     1       /(2.*A(SIG,THETA)+F(X,THETA)**2))
C
C     WRITE(*,*)'debug: AAX=',AAX
      END FUNCTION AAX
C
C-------
C     FUNCTION APHI(X,SIG,THETA)
C
C     COMMON PI,B,LAMBDA,GZ,vratio,etadr
C
C     REAL LAMBDA
C
      APHI = SIG/(4.*F(X,THETA)*COS(THETA) - SIG)
C
      IF(APHI.LT.0.)WRITE(21,*)'WARNING:APHI:',APHI
      END FUNCTION APHI
C
C-------
C     FUNCTION A(SI,TH)
C
C     COMMON PI,B,LAMBDA,GZ,vratio,etadr
C
C     REAL LAMBDA
C
      A  = SI* COS(TH)/(4.*SIN(TH)*SIN(TH))
C
      IF(A.LT.0)WRITE(21,*)'WARNING:A=',A
      END FUNCTION A
C
C-------
C     FUNCTION F(X,THETA)
C
C     Tip-Loss model from L. Prandtl (1919)
C
C     COMMON PI,B,LAMBDA,GZ,vratio,etadr
C
C     REAL LAMBDA
C
      call shen(gc)
      F = gc*.5*B*(1.-X)/(X*SIN(THETA))
C
      F = (2./PI) * ACOS(EXP (-F))
C
C     WRITE(*,*)'debug: F=',F
      END FUNCTION F
C
C-------
```

RA.f

Sep 05, 13 14:15	RA.f

```
c
      sth  = sin(theta)
      cth  = cos(theta)
      cth  = cth/sth
      la   = LAMBDA
      ftip = F(x,theta)
      AA   = AAX(X,SIGMA,THETA)*ftip
c
      dcp = aa*(1.-aa)*(tan(theta)    -drag)*x*x*dx
      dct = aa*(1.-aa)*(1.+tan(theta)*drag)  *x*dx
c
      cptot = cptot + dcp
      cttot = cttot + dct
C-----
C     output   localen values
c
      WRITE(IOOUT1,'(7F12.4)')X,xx,CCLOUT,1.e3*X*Rtip,
     +       1.e3*CCLOUT*Rtip/Cldes,
     +       180.*THETA/PI,ftip
c
      WRITE(*       ,'(7F12.4)')x,xx,CCLOUT,1.e3*x*Rtip,
     +       1.e3*CCLOUT*Rtip/Cldes,
     +       180.*THETA/PI,ftip
c
      WRITE(IOOUT2,'(5F12.6)')X, AXOUT,PHIOUT, dcp,AA
c
c     output for blade desing (CAD) tool
c
      WRITE(IOOUT3,'(aI0,FI0.4,fI0.1,2FI0.4i5)')
     +       steuer,X*Rtip.180*theta/pi,cclout*rtip,aec,ipr
C----
      END X loop = 1, nx
c
      END DO
c
c     this normalization seems highly unclear !!!
c
      cptot = 8.*la*cptot
      cttot = 8.*cttot
      cpauto = etadr*(1+1./vratio)*cptot - cttot
      aa   = 0.5*(1.-sqrt(1.-cttot))
c
      WRITE(*,*)
      WRITE(*,'(a4,2a8, 4a11 )')'B','LAMBDA','GZ','cP',
     +       'cT','cP-car/ea','a'
      WRITE(* ,'(i4,2f8.1,5f9.3)') int(B) , LAMBDA, GZ,
     +                    cptot,cttot,cpauto,aa
c
C-------------------------------------------------
c     WRITE(ioout1, '(a4,a8,   5a9   )')'B','LAMBDA','GZ','a'
c     WRITE(ioout1, '(a4,2a8, 5a9  )')'cP','cT','cP-car/ea','a'
c     WRITE(ioout1 ,'(i4,2f8.1,5f9.3)') int(B) , LAMBDA, GZ,
c     +                    cptot,cttot,cpauto,aa
C-------------------------------------------------
c
      CLOSE(IOOUT1)
      CLOSE(IOOUT2)
c
      END
C
C-------
C-------
C-------
```

Seite 7/8

Sep 05, 13 14:15 — RA.f

```fortran
      SUBROUTINE shen(gc)
C
C     tip loss model from soerensen, shen et al (2004)
C
      COMMON PI, B, LAMBDA, GZ, vratio, etadr
C
      REAL LAMBDA, gc
C shen
C
      gc = 0.1 + exp(-0.125*(B*lambda-21.))
C kein shen
C
C     gc = 1.
      END
C
      FUNCTION CCL(X,SIG,THETA)
C
      COMMON PI,B,LAMBDA,GZ,vratio,etadr
C
      REAL LAMBDA
C
      CCL = 2.*PI*X*SIG/B
C     WRITE(*,*)'CCL=',CCL
      END FUNCTION CCL
C
      FUNCTION GRX(X,SIG,THETA)
C
      COMMON PI,B,LAMBDA,GZ,vratio,etadr
C
      REAL LAMBDA
C
      GRX=LAMBDA * X -
     +  AAX(X,SIG,THETA) * (1.-AAX(X,SIG,THETA))*F(X,THETA)*TAN(THETA)
     +  /(APHI(X,SIG,THETA)*(1.-AAX(X,SIG,THETA)))
C
C     WRITE(21,*)'GRX=',GRX
C     WRITE(21,*)'L,X,AAX,TH,APHI',LAMBDA,X,AAX(X,SIG,THETA),THETA,
     +                              APHI(X,SIG,THETA)
      END FUNCTION GRX
C
      FUNCTION DELTACPauto(X,S,T)
C
      COMMON PI,B,LAMBDA,GZ,vratio,etadr
C
      REAL LAMBDA
C
      DELTACPauto = etadr*(1.+1./vratio)*deltacp(x,s,t)-deltact(x,s,t)
C
      END FUNCTION DELTACPauto
C
      FUNCTION DELTACT(X,S,T)
```

Seite 8/8

Sep 05, 13 14:15 — RA.f

```fortran
C
      COMMON PI,B,LAMBDA,GZ,vratio,etadr
C
      REAL LAMBDA
C
      xx = lambda*x
C
      DELTACT = (TAN(T)/GZ + 1.)
     +    *    AAX(X,S,T)*F(X,T)
     +    * (1.-AAX(X,S,T)*F(X,T))/xx
C
      END FUNCTION DELTACT
C
      FUNCTION DELTACP(X,S,T)
C
C     formel (6) auf seite 4-17
C
      COMMON PI,B,LAMBDA,GZ,vratio,etadr
C
      REAL LAMBDA
C
      DELTACP = (-1./GZ+TAN(T))
     +    *    AAX(X,S,T)*F(X,T)
     +    * (1.-AAX(X,S,T) F(X,T))
C
C     WRITE(*,*)'debug; DELTACP= ',DELTACP
      END FUNCTION DELTACP
C
      FUNCTION DGDTHETA(X,S,T)
C
      COMMON PI,B,LAMBDA,GZ,vratio,etadr
C
      DT = T/100.
      DGDTHETA = (GRX(X,S,T+DT)-GRX(X,S,T))/DT
C     WRITE(21,*)'DGDTHETA=',DGDTHETA
      END FUNCTION DGDTHETA
C
      FUNCTION DGDSIGMA(X,S,T)
C
      COMMON PI,B,LAMBDA,GZ,vratio,etadr
C
      DS = S/100.
      DGDSIGMA = (GRX(X,S+DS,T)-GRX(X,S,T))/DS
C     WRITE(21,*)'DGDSIGMA=',DGDSIGMA
      END FUNCTION DGDSIGMA
```

References

1. Batchelor GK (2000) An introduction to fluid dynamics. Cambridge University Press, Cambridge
2. Betz A (1919) Schraubenpropeller mit geringstem Energieverlust. Nachrichten der Königlichen Gesellschaft der Wissenschaften zu Göttingen, pp 193–217
3. Betz A (1964) Konforme abbildung. Springer, Berlin
4. Branlard E, Dixon K, Gaunaa M (2012) An improved tip-loss correction based on vortex code results. In: Proceedings of EWEA 2012, Copenhagen, Denmark
5. Glauert H (1934) Aerodynamic theory, vol IV. Springer, Berlin, pp 261 ff
6. Goldstein S (1929) On the vortex theory of screw propellers. Proc Roy Soc 123A:440–463
7. Gontier H et al (2007) A comparison of fatigue loads of wind turbine resulting from a non-Gaussian turbulence model vs. standard ones. J Phys: Conf Ser 75:012070
8. Hansen MOL, Johansen J (2004) Tip studies using CFD and comparison with tip loss models. In: Proceedings of 1st conference on the science of making Torque from wind, Delft, The Netherlands
9. van Kuik G (2013) On the generation of vorticity in rotor & disk flows. In: Proceedings of ICOWES2013, Copenhagen, Denmark
10. Lang S (2000) Complex analysis, 4th edn. Springer, New York
11. Mann J (1994) The spatial structure of neutral atmospheric surface-layer turbulence. J Fluid Mech 273:141–168
12. Mann J (1998) Wind field simulation. Prob Eng Mech 13, 4:269–282
13. Okulov V, Sørensen JN (2008) Refined Betz limit for rotors with a finite number of blades. Wind Energy 11:415–426
14. Philip J (2013) Spatial averaging of velocity measurements in wall-bounded turbulence: single hot wires. Meas Sci Technol 24:115301
15. Renner C (2002) Markowanalysen stochastischer fluktuierender Zeitserien, Dissertation, Carl von Ossietzky Universität Oldenburg (in German)
16. Schepers G (2013) Private communication, 3rd IEAWind Task 29 (MexNext II) meeting, Pamplona, Spain. Accessed 27 Sept 2013
17. Shen WZ, Mikkelsen R, Sørensen JN (2005) Tip loss corrections for wind turbine computations. Wind Energy 4:457–475

About the Author

A. P. Schaffarczyk spent his childhood gazing at the blue sky of Northwestern Germany (*Star-fighters* flashing by) and also watching astonished on black-and-white TV the rocket *Saturn V* launches that brought the first men to the moon in the 1960s.

Interested in mathematics and physics, he attended University Göttingen, well-known for mathematics and physics but also for Ludwig Prandtl's *Aerodynamische Versuchsanstalt*. After receiving his Ph.D. in some kind of Statistical Mechanics, he came into deeper touch with fluid mechanics after his employment in industry. There he learned to apply all his learned theoretical knowledge and skills to solve practical problems. In 1992 he joined *University of Applied Sciences Kiel, Germany* and worked further on computational fluid dynamics and, since 1997, on the aerodynamics of wind turbines.

© Springer Nature Switzerland AG 2020
A. P. Schaffarczyk, *Introduction to Wind Turbine Aerodynamics*,
Green Energy and Technology, https://doi.org/10.1007/978-3-030-41028-5

Glossary

ABL   Atmospheric boundary layer

AOA   Angle-Of-Attack: angle between chord line (longest line between nose and tail of an aerodynamic profile) and inflow wind

Boundary layer   A concept introduced by Ludwig Prandtl in which the influence of viscosity in high Reynolds number flow is confined in a thin layer close to the wall and the body

GROWIAN   **GRO**sse **WI**nd **AN**lage = large wind turbine

Hub   Height of main shaft

Ising model   Model system in statistical physics

Kölner KK   Kölner Kryo Kanal = Cryogenic wind tunnel at Cologne, Germany. Here Reynolds number is increased by cooling the tunnel gas (nitrogen) down to $-200°$ K

Tensor   In most cases represented by a matrix they are defined mathematically as linear relations between vectors

Vortex generators   Small upright oriented triangles which shed small vortices into the boundary layer and may prevent early stall

Windmill   A Windmill is an early form of a *working machine* which was (is) used for grinding wheat.

Wind index   A positive number which correlates the annual average wind speed to a long-term average.

Wind turbine   A Wind turbine is an engine which converts kinetic energy from the wind, mostly into electricity.

© Springer Nature Switzerland AG 2020
A. P. Schaffarczyk, *Introduction to Wind Turbine Aerodynamics*,
Green Energy and Technology, https://doi.org/10.1007/978-3-030-41028-5

Index

© Springer Nature Switzerland AG 2020
A. P. Schaffarczyk, *Introduction to Wind Turbine Aerodynamics*,
Green Energy and Technology, https://doi.org/10.1007/978-3-030-41028-5

Printed in the United States
By Bookmasters